**Longman Handbooks in Agriculture**

*Series editors*
C. T. Whittemore
R. J. Thomas
J. H. D. Prescott

Whittemore: *Lactation of the Dairy Cow*
Whittemore: *Pig Production: the Scientific and Practical Principles*
Speedy: *Sheep Production: Science into Practice*

# Lactation of the dairy cow

**Colin T. Whittemore**

*Head of Animal Production Advisory and Development, Edinburgh School of Agriculture*

**Longman** London and New York

To my wife

**Longman Group Limited** London

*Associated companies, branches and representatives throughout the world*

*Published in the United States of America by Longman Inc., New York*



*First published 1980*

---

**British Library Cataloguing in Publication Data**
Whittemore, Colin Trengove
Lactation of the dairy cow. – (Longman handbooks in agriculture).
1. Milk production 2. Lactation
I. Title
636.2′1′4 SF208 79-40442

ISBN 0-582-45079-9

---

Printed in Great Britain by McCorquodale (Newton) Ltd

# Contents

*Preface*

**1 The milk industry 1**

Breeds and yields *1*
Use *3*

**2 The mammary glands 5**

Mammary growth *9*
Structure of the lactating mammary gland *13*
Secretion *19*
Hormones involved in lactation *22*
Milk ejection *26*

**3 Natural milking 29**

Initiation *33*
Suckling *34*
  Passive withdrawal *34*
  Milk ejection *35*
  Termination *39*
Nursing and sucking behaviour *39*

**4 Machine milking 42**

Machine pulsation *47*
Stimulation *49*
Stripping *52*
Mastitis *55*

Frequency of milking *56*
Parlour routine *56*

**5 Characteristics of lactation yield 63**

Seasonality *66*
The lactation curve *67*
Nutrition *71*
Lactation length *72*

**6 Nutritional value and quality of milk 76**

Nutritive content *77*
Factors affecting compositional quality *80*
Nutrition and milk composition *83*
Payments for quantity and quality *85*

*Tailpiece 87*
*Further reading 89*
*Index 90*

# Preface

Milk production is a business with profit as the motive. But feeding people, particularly with cow's milk, is also a *responsibility* of the agricultural community. To increase the efficiency of milk production, there must be some understanding of that fascinating female – the dairy cow. The ecological pressures of the modern world make it essential that food production by use of animals be no lackadaisical or rustic affair if resources are not to be squandered. Milk can no longer be properly produced on the basis of hearsay and the experience of accepted practice; apprenticeship by example is not good enough. This is not to say that the fund of knowledge built up from years of experience is necessarily wrong; but it is to say that explanations for natural responses are the better for being founded in logic or in science. The aim of what follows is therefore to provide a link between the science and the practice of lactation; and on the way perhaps to throw some light on what makes the dairy cow tick.

In the preparation of the book I have received help from so many friends and colleagues, particularly at the Edinburgh School of Agriculture, that I cannot thank them all individually. I would also like to acknowledge the great fund of learning, compiled at the Universities and Research Institutes, from which I have drawn.

The photographs were taken at Langhill by Mr. G Finnie.

CTW
*Edinburgh 1979*

# The milk industry

1

Milk is extracted from the dairy cow for feeding to people. But in all but totally agrarian communities it is necessary for a preparative and distribution industry to stand between the milk producer and customer, and sell liquid milk or milk products to the requisite outlets. This is not about that industry, nor about the populations it serves, but it is salutory for all concerned with the dairy cow briefly to consider what befalls her efforts on our behalf.

## Breeds and yields

Milk production from some countries around the world is shown in Table 1.1. Goat and sheep milk comprise a very small proportion of the total. Of the cow milk, most comes from the European cattle breeds. In Asia milk production is divided between the Zebu cattle (for example, the Sahiwal and Sindhi) and buffalo (for example, the Murrah). The European breeds can yield around 4000–6000 kg of milk per lactation of 300–350 days, while the Zebu and buffalo tend to average somewhat less than 2500 kg in a shorter lactation of 250 days or so. Conditions of husbandry differ widely between European cattle and other cow types, and the former do not always shape up well in tropical situations; however, the crossbred Zebu/Friesian can sometimes double the yield of native breeds.

The dairy cattle populations of a few Western countries with

**Table 1.1** Milk production (million tonnes)

| | Cow | Goat* | Sheep† | Estimated cow yield (t/yr) |
|---|---|---|---|---|
| West Germany | 21 | 0.02 | — | 4.0 |
| France | 29 | 0.30 | 0.80 | 3.2 |
| Netherlands | 9 | — | — | 4.5 |
| United Kingdom | 14 | — | — | 4.2 |
| USA | 52 | — | — | 4.5 |
| USSR | 87 | — | — | 2.5 |
| Asia | 46‡ | 3 | 3 | 1.5 |

*Goats are important milk producers within extensive agricultural systems. The potential yield of a 40–60 kg goat is 800–1000 kg milk in 300 days or so. In comparison, the conventional cow of ten times that weight only yields four times the milk.
†The best milk sheep is the East Friesian, which can produce 300–500 kg per lactation.
‡Some 20 million tonnes of this is estimated to be buffalo milk. Buffalo can yield up to 4 t/year with an average of 1–2 t.

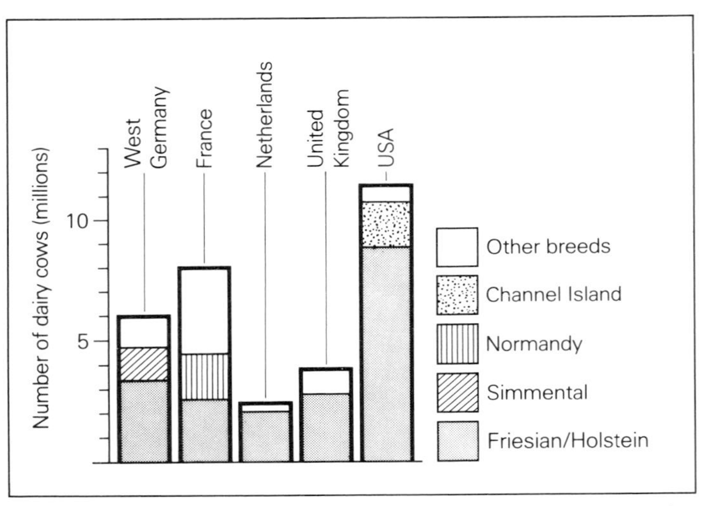


**Figure 1.1** Numbers of cows in some countries together with the main breeds

advanced dairy industries are depicted in Fig. 1.1. With the exception of France, the Friesian breed reigns supreme in each of these countries. The average herd size in West Germany and France is about 20 cows, but average lactation yield is higher in Germany. The highest yields yields per cow ($4\frac{1}{2}$ tonnes/lactation) come from the Netherlands, whose average herd size is 40. The UK has by far the largest number of big herds, with 70 being the average and 100–200 cow herds common. Individual cow yield in the UK is about $4\frac{1}{4}$ t/lactation. The trend to increased individual cow yield is likely to continue as our knowledge of how to care for dairy cows increases, and allows actual yields (currently about 4000 kg) to more nearly approach the potential (probably about 8000 kg). Likewise herds will grow larger than the present UK average of about 70 just so long as the economies of scale in buildings, milking parlours, equipment and men continue to operate in favour of herd units of 200 or more cows.

## Use

The utilization of milk on the European continent (Fig. 1.2) shows a consistent pattern: about 20 per cent of milk is drunk as a liquid, about 40 per cent goes to butter manufacture and 15 per cent or so to cheese. The UK and the USA differ in consuming a much higher proportion (about 50%) of their milk in the liquid form, and diverting relatively less to manufacture

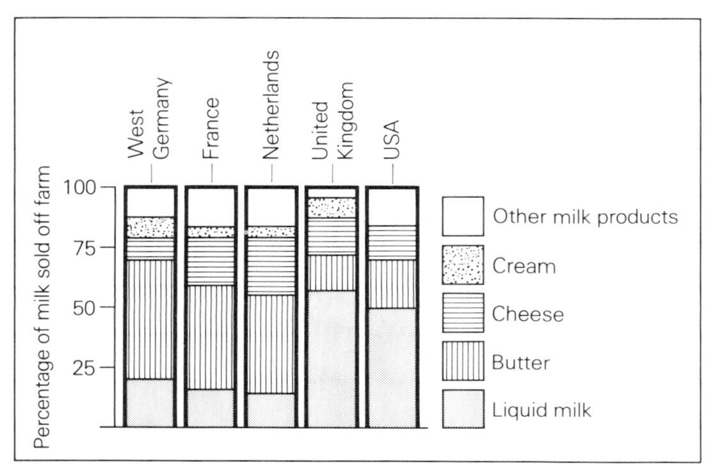


**Figure 1.2** Utilization of milk sold off farms

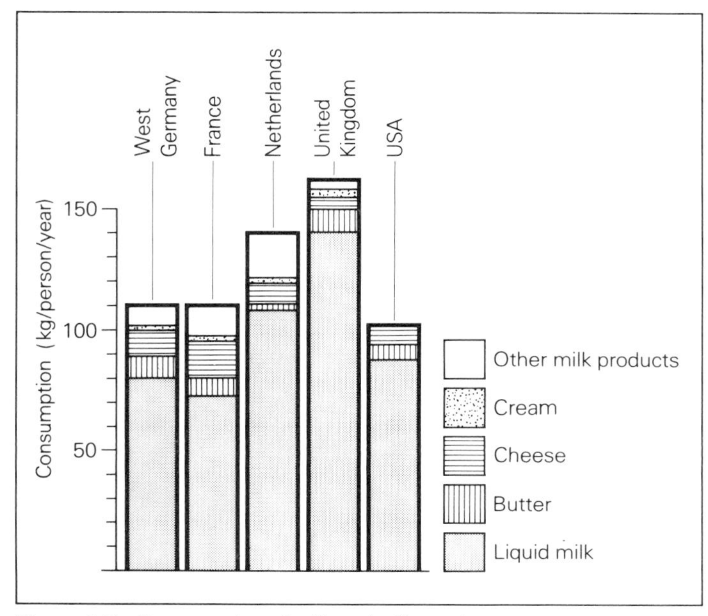


**Figure 1.3** Consumption of milk and milk products

into milk products. The patterns of consumption of milk and milk products (Fig. 1.3) do not mirror the patterns of utilization, and there is therefore trade between the countries in milk products (considerable quantities of German, French or Dutch butter, for example, are eaten in the UK). The Germans drink less milk than the Dutch, while the British drink almost twice as much milk as the French. Great Britain is less than 60 per cent self-sufficient in milk fat, and in the context of island self-sufficiency there could be something to be said for increasing the production of milk fat.

By the consumption of 100–150 kg of milk and milk products per person per year, the Western world bears ample evidence to the value it places on cow's milk both as a source of nutrients, a convenience food and a luxury. It is perhaps also worth observing that in real value terms the price of milk has consistently come down over the last 40 years; a tribute to steps already made in enhancing the efficiency of intensive milk production.

# The mammary glands

# 2

Milk is manufactured from simple blood nutrients by the milk-synthesizing cells of special glands, the mammary glands. The mammary glands, or mammae, are skin glands, albeit large ones, held exterior to the body cavity. Mammary tissue therefore forgoes the potential advantage of rigid skeletal support. This brings its own problems for modern dairy cows which have been selected, and are farmed, to produce much greater quantities of milk than the original structure was designed to cope with. The mammae are soft and tend to be pendulous. The consequences for the shape and structural integrity of the mammary glands which follow from the fact that they do, perforce, hang from the body rather than being contained within it, are confirmed by observation of the ungainly appearance, when running, of any female mammal in full lactation.

The pliable nature of mammary tissue follows from its construction. During lactation, much of the mass is comprised of the fluid milk itself together with the delicate tissue which secreted it, and the ducting systems required to lead milk from the point of secretion by individual cells to the nipple or teat. When not in lactation the mamma is smaller in size, though not in proportion to the absence of the secretory tissue and its milk products, for the space vacated is replaced to some extent by connective and fatty tissue. Shapeliness of the virgin mammary gland may often be a function of the degree of invasion by fatty cells, and therefore does not necessarily bear any relationship to future milk production capacity.

**Figure 2.1.** External appearance of mammary glands of dairy cow

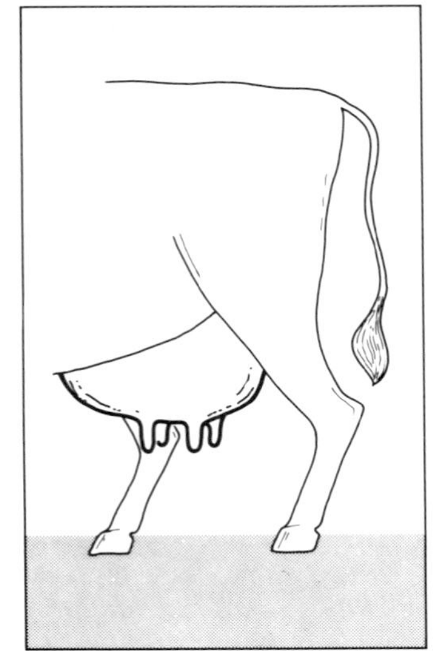

(*a*)

Four mammary glands comprising an udder of the type that milking machine teat cups were designed for: the teats are evenly spaced, of equal length, straight sided, at right-angles to the surface of the mammae and held well clear of the ground

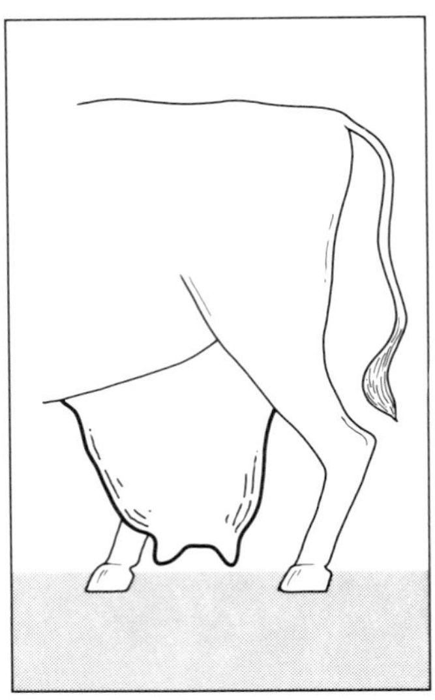

(*b*)

Inadequate ground clearance resulting from structural problems: teats prone to contamination and damage; teat cups of machine difficult to apply. The illustration also shows short, fat and triangular teats which are not readily drawn in, or held, by the teat cups

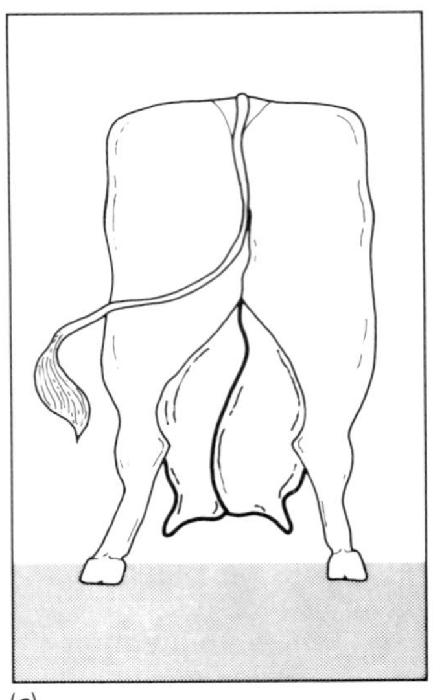

(*c*)

Teats sticking-out sideways: ground clearance improved, but distance between teat tips presents the operative who has to attach teat cups with problems. There is also increased risk of machine falling off due to vacuum escape

For its support, the mammary gland must rely on sheets of relatively soft connective tissue, together with, in the dairy cow, some slightly more robust suspensory ligaments which form the main, centrally placed, attachment of the mammae to the body wall. The mammary tissue is maintained quite separately from other body tissue, and the supply of blood vessels and nerves to the gland is confined to relatively few discrete groupings; the majority passing from the body cavity through a single passage in the body wall to the tissues of the gland on the body exterior. The manipulability of the shape of the mamma bears evidence to the relative frailty of its supportive tissues: the brassière industry is inventive enough to allow the phenomenon of fashion to govern both the currently preferred shape and the position of human mammary glands. The same inherent weakness of the suspensory ligaments to cope with the loading placed upon them by the high-yielding dairy cow make the often mooted possibility of foundation garments for cows not an entirely fanciful concept.

Like any well-designed manufacturing plant, the function of the mammary gland should govern its structure, and structure should enable maximum production efficiency. Thus, mammae with high milk secretory activity but with inadequate capacity to store and carry milk between milkings are unacceptable in the context of modern milk production and twice daily milking, although quite acceptable for a sucking calf with freedom of access to the mammae for more frequent milk removal. Again, the gross stretching which can occur in glands with inadequate support capability in relation to productive loading, may, by awkward shaping and positioning of the udder, make milk removal by machine unduly time consuming and complex for the requirements of an efficient milking parlour routine. Similarly, teat shape and their position on the udder become relevant because of the inflexibility of the milking machine as compared to the agility of a hungry calf (see Fig. 2.1). Whilst there must be a *general* positive relationship between yield and gland size and shape, this does not hold for all *particular* circumstances. The udder shown in Fig. 2.2(a) yielded twice the amount of Fig. 2.2(b).

The major function of the mammae must be to produce copious quantities of milk. For this, first the digestive tract, and then the gland via the blood circulation system must receive even more copious quantities of the energy and protein-containing constituents needed as raw materials for milk manufacture (synthesis). The synthesizing cells must themselves not only be present in great number, but must also be highly active; the rate of milk production being the product of cell number multiplied by cell activity.

The ultimate activity of the mamma is to readily and fully release its contents to the outside world. If the secretory products are not to be squandered, the gland must effectively store and retain the milk it has made until it is needed; the mechanisms for storage and retention must thus be flung into reverse when suckling, or milking, time arrives. It is essential

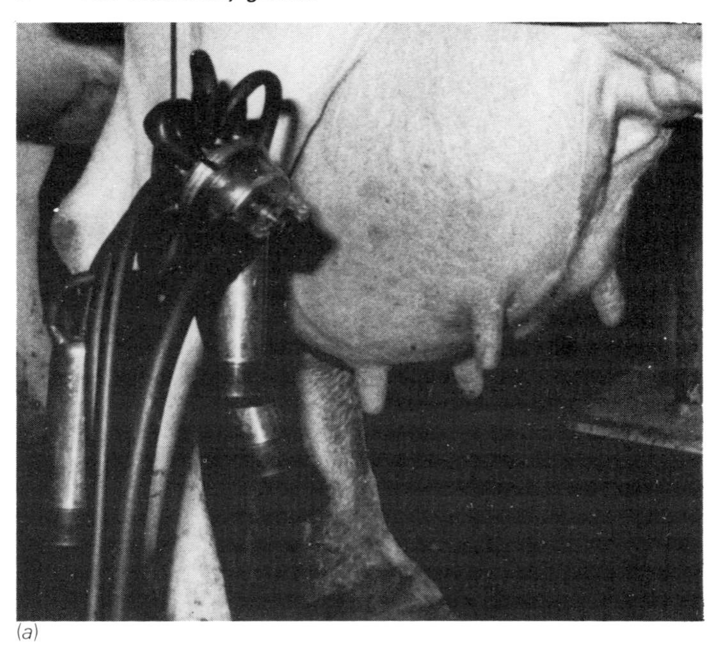

(*a*)

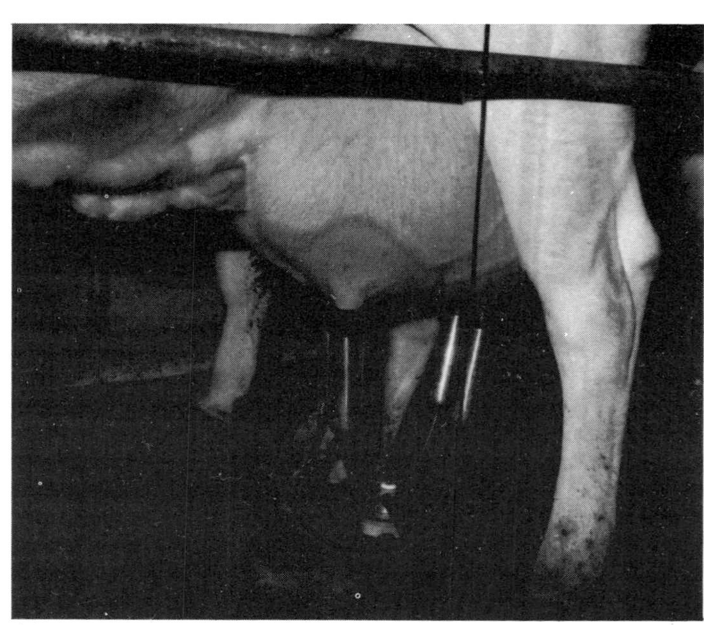

(*b*)

**Figure 2.2** Appearances can sometimes deceive: cow (*a*) far outyielded cow (*b*)

that milk release, when it does occur, is not haphazard such as would result from a mechanism triggered off by gland fullness or the production of a given amount of milk. Milk must flow only when an appropriate receptacle for its receipt is attached to the teat, be it the teat cup of a machine or the mouth of her young. There must be further safeguards to ensure fair shares for all in the case of animals who may give birth to more than one offspring. Mechanisms for milk retention and milk release, while complex and incompletely known, are, by and large, efficient. The interaction between animal physiology and animal psychology has always been at the heart of the dairy parlour routine, but in the first place the animal must be willing to be milked by a machine rather than by her offspring. In the natural situation, not only is a calf the only means of milk removal, but it must be one particular calf which has the unique right of ownership to the products of his dam's mammary glands. The pig takes this particular characteristic to the ultimate extreme whereby each piglet of the litter owns an individual nipple. The necessarily common characteristics of species which produce milk for human consumption is that they will readily allow themselves to be milked. In contrast to 'European' breeds of cattle, the Zebu and buffalo types tend to be reluctant in this respect. This problem is particularly relevant when it is realized that these two latter cow types are adapted to those particular climates where the human population has nutritional problems that could, in large part, be overcome by an increased consumption of milk.

At the present moment there is no doubt that the potential for milk production is considerably above current performance levels. Consideration to the principles which govern the functioning of the mammary glands of the dairy cow may go some way to narrowing the gap between average and possible levels of milk yield.

## Mammary growth

The embryonic mammary gland is initiated as a mammary bud (Fig. 2.3*a*). Canals form, and these are antecedents of the streak canal, and teat and gland cisterns (Fig. 2.3*b*). Next, the multi-directional system of milk passages is formed, and comprises sinuses and large and small ducts. The sinuses are extensions of the cistern and, like the cistern, serve as milk storage vessels. The ducts are used to lead milk down from the parts of the gland where it has been manufactured into the sinuses and cistern. By the time of birth, a rudimentary teat, together with a rudimentary duct system, is formed in both sexes (Fig. 2.3*c*). There is, as yet, no secretory tissue differentiated, and, in common with the rest of the body at birth, little or no fatty tissue.

After birth the duct system continues to extend and branch to ever greater extremes of fineness, while all the elements of the gland grow in size and accumulate fat and connective tissue in pace with the normal growth of the general body

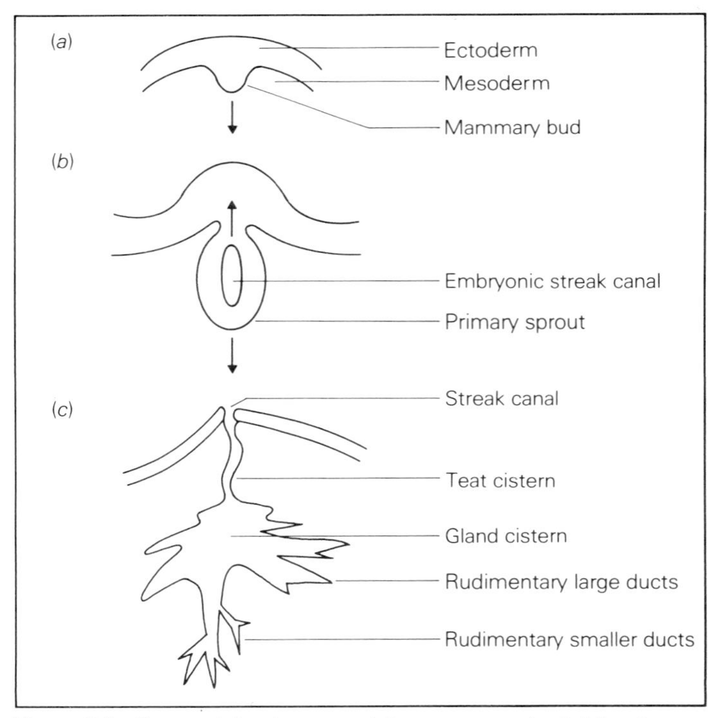


**Figure 2.3** Pre-natal development of the mammary gland: (*a*) and (*b*) embryonic development; (*c*) extent of development by birth

mass. Between 3 and 9 months of age the cells of the mammary gland in the young female cow, but not in the young male, multiply very much more rapidly than those of the rest of the body, causing a relative growth spurt during which about one-tenth of the potential total growth is made and the glands become more readily visible.

Puberty for the dairy heifer is in part dependant upon body weight, occurring at around 260 kg, and also upon body condition (fatness), breed and season. Puberty arrives at about 9–10 months of age under European conditions and brings with it the hormonal changes associated with the reproductive (oestrus) cycle. At each cycle of about 3 weeks, oestrogen hormones encourage proliferation of the ducting system, but this is followed closely by its regression. The flow and ebb of tissue generation and degeneration is balanced slightly in favour of positive growth so that a highly branched system of milk passages gradually develops within the mamma. There is, as yet, no specific development of the milk-secreting alveoli. The mammae of a dairy heifer may grow by about 200 grams per month between 9 and 13 months of age. Because the effective gain from growth waves with oestrus cycles is relatively small, heifers are likely to have little more useful mammary tissue at 20 months of age than they had at 13 months. The possibility that a forthcoming first lactation might be enhanced by delaying the mating of a well grown female is therefore remote.

In the absence of conception, the post-pubertal growth

surge leaves the developed mamma at a size which although fundamentally related to body mass is greatly modified by inherited characteristics and by nutritional supply. The mammary gland at this stage has a nipple with a ring (the teat areola) of touch-sensitive tissue. In the young dairy cow, each nipple has for its orifice a single 'streak' canal leading out from the teat cistern. The pig has two such canals exiting on a flattened plane just below the tip of the nipple, while the human female may have two, three or more canals at the nipple end, and indeed, occasionally small canals elsewhere in the area of the nipple through which milk may be seen to exude in early lactation. The internal structure of the virgin gland has a complete, though poorly developed, blood and nerve supply system, and the gland cistern, sinuses, large ducts, smaller ducts and finally the fine ducts are all functional. Nevertheless, the majority of the mamma is fatty tissue together with connective tissue of undifferentiated body cells and some structural collagen. The undifferentiated cells are destined to develop into the active milk secretory cells with their support tissues when, in the fullness of time, pregnancy intercedes.

The purpose of milk is to feed the newborn young; it is hardly surprising therefore that it is pregnancy which is responsible for initiating the milk-manufacturing process. With the onset of pregnancy, the internal elements of the mammary gland are frenetically active. It is during pregnancy that by far the greatest development takes place, bringing about a final structure complete by full term. In the first 3 months, under the influence of the pregnancy hormones (particularly progesterone and prolactin), the ducts proliferate further into the undifferentiated tissue mass. During mid-pregnancy this mass itself begins to differentiate into lobes (clusters) of alveolar tissue which is made up of milk-secreting cells arranged around the inside of microscopic spheres (alveoli). From the sixth month of pregnancy, lobes of alveolar tissue are clearly defined, and during the next 3 months the alveoli complete their formation and fill with a syrupy secretion. The gland becomes less malleable, more ducts and secretory tissue replace fat, the mass weight increases as does the volume, the skin stretches, and there are sensations of pressure, tenderness and heat. During the last third of pregnancy most of the development will have been of the active layer (epithelium) of milk-manufacturing cells which lie on the inner surface of the newly formed alveoli. Over the final month, secretory activity is accelerated by a rapid increase in the number of cells and their synthesis rate. Within 7 days of the impending birth, the mamma is full of milk, and only back pressure (because there is no milk being withdrawn) is preventing the secretory cells from continuous synthesis of milk. As a result of the very rapid development of the active mammary tissue in these last stages of growth, the major part of the total secretory tissue present at the beginning of lactation is formed in a relatively small proportion of the total development phase (Table 2.1).

**Table 2.1** Possible number of cells in the mammary gland, expressed as a percentage of the maximum number of cells present at peak yield.

| | Percentage of cell number at peak yield |
|---|---|
| Birth | <1 |
| Puberty (10 months) | 5 |
| Mating (15 months) | 10 |
| Half-term pregnancy (20 months) | 20 |
| Full-term pregnancy (24 months) | 60 |
| Peak lactation (26 months) | 100 |

The removal of milk from the gland, usually by the newborn baby but not necessarily so, signals the initiation of the lactation. Mammary growth continues into early lactation, and a significant increase in secretory tissue occurs. During the immediate post-natal period tissue development proceeds in proportion to the positive stimulus received by the frequency and completeness of milk removal. In some cases, the increase in secretory cell number after birth may equal that achieved before birth (Fig. 2.4).

It has been estimated that the dairy cow's mammary glands probably contain in the region of $5\times10^{12}$ secretory cells in the epithelium of alveolar tissue. Like all active body cells, these have a short life and are rapidly turned over. In early lactation replacement by the manufacture of new cells greatly exceeds the rate of loss, and there is a net gain. After peak yield has been reached, the balance of gain and loss is tipped in the favour of a gradual decline in total cell number. In the dairy cow this occurs from the second to third month or so after calving, and from that time on the lactation is inescapably programmed into decline towards the end of the lactation when the active secretory cells of the alveoli regress and degenerate.

For dairy cows mated within 3 months of parturition, the eighth month of the lactation has particular significance because it coincides with the fifth month of pregnancy. The hormones of pregnancy cause an increase in the rate of decline both of mammary cell number and also of the synthetic activity of individual cells. It can be no accident that the impending arrival of a new addition should bring about a cut-back and final termination of mother's food supply to the current offspring. Not only is it inconceivable that a boisterous one-year-old would willingly share his source of ready food and comfort with a new-born interloper, but the mammary tissue must itself be regenerated.

Some dairy cows may continue to produce milk right up to the birth of the next calf. As this would disallow the essential regeneration phase, such animals need to be forcefully dried off by stopping milking at least 1 month before the next birth is due. Following cessation of milk removal, there is a build-up of milk in the mamma before synthesis and secretion stop. The extent of the build-up depends on the level of activity in the secretory epithelium when milk removal ceases, and is clearly greater in dairy cows still yielding significant quantities

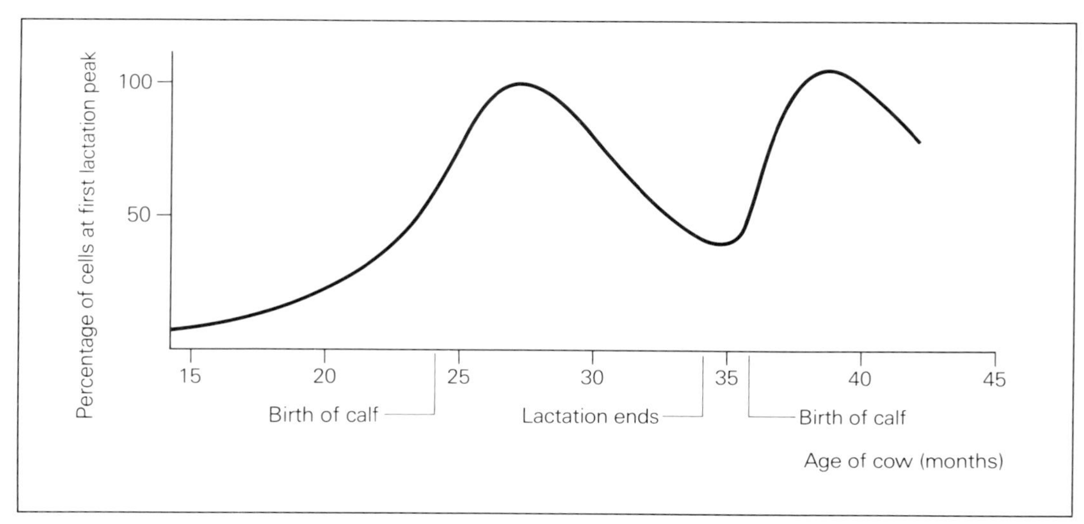


**Figure 2.4** Changes in number of milk manufacturing cells in the mammary glands of a dairy cow (expressed as a percentage of the cells present at first peak)

of milk even after 10 months of lactation. Termination of milk synthesis by the cells is quickly followed by degeneration of the active secretory layer of alveolar epithelium. The remainder of the alveolus and duct structure remains intact. After a brief respite, the maternal body reawakens to the inevitable approach of the next birth some 3 to 4 weeks hence, and a new layer of secretory epithelial cells is formed within the alveolar framework. This comprises the beginning of the next acceleratory growth phase for the gland as it prepares for the next lactation.

## Structure of the lactating mammary gland

The cow carries four mammary glands, the rear two accounting for about 60 per cent of total milk production. The human carries two mammae and the pig 10 to 12 or more. In the case of the cow the only visible evidence of a quartet of glands are the teats, and some cleavage down the mid line evidencing the support role of the medial suspensory ligament (Fig. 2.5). The glands of the cow are formed into the udder, the tissues of which may weigh 15–30 kg and which may hold the same weight again of milk. Limits to the rate of milk production are a combined function of the mass of tissue actively synthesizing milk together with the capacity of the cisterns, sinuses, ducts and internal alveolar spaces (lumena) to store the milk once secreted. Secretory products from the epithelial cells first fill the lumen of the alveolus, and then under the pressure of gravity and continued synthesis, the newly formed milk passes down the ducting system. Immediately prior to milking about 40–50 per cent of the total milk is held in the gland cisterns, sinuses and large ducts; while the other 50 per cent is held in small ducts and alveolar lumena.

The udder of the cow is supported from the body wall and pelvic girdle by suspensory ligaments. The lateral suspensory ligaments (see Fig. 2.6*a*, page 17) are fibrous while the medial suspensory ligaments (Fig. 2.5) are more elastic. As the mammae fill with milk the udder drops and stretches; udder stretching allows for accommodation of about 50 per cent of the half-daily yield. It is the greater elasticity of the medial ligaments which can cause the bizarre appearance of some high-yielding cows awaiting milking. Not only may udder stretching force her hind legs apart to make walking difficult, but the middle of the udder may drop further than the outside, causing the milk-engorged teats to project outwards rather than to hang down from the udder.

The teat orifice is closed by a small but strong sphincter muscle (Fig. 2.5), behind which lies 8–12 mm of streak canal. The teat cistern holds only 30–50 ml of milk, and between the teat cistern and the larger gland cistern there are annular folds of tissue which serve to retain milk in the gland cistern and reduce pressure on the sphincter at the teat end. The size of the gland cistern differs greatly between species, that of the pig being characteristically small, and of the dairy cow characteristically large, holding 500–2000 ml. The duct system leads into the cistern through 8–12 large portals, while the terminal parts of the ducts comprise many large spaces or sinuses. In terms of function the cistern, sinuses and large ducts should perhaps not be differentiated, all being inactive with regard to milk production and used only for storage. The duct system terminates at the functioning milk-manufacturing unit, which is a microscopic lobule of about 200 alveolar cells occupying a fraction of a cubic millimetre (Fig. 2.5). From the lobules the small ducts gather together into larger interlobar ducts leading milk from groups of lobules, or lobes. The interlobar ducts gather to form the large ducts and milk sinuses which gather at the gland cistern. Throughout the mammae, lobules and lobes are separated by septa of connective tissue.

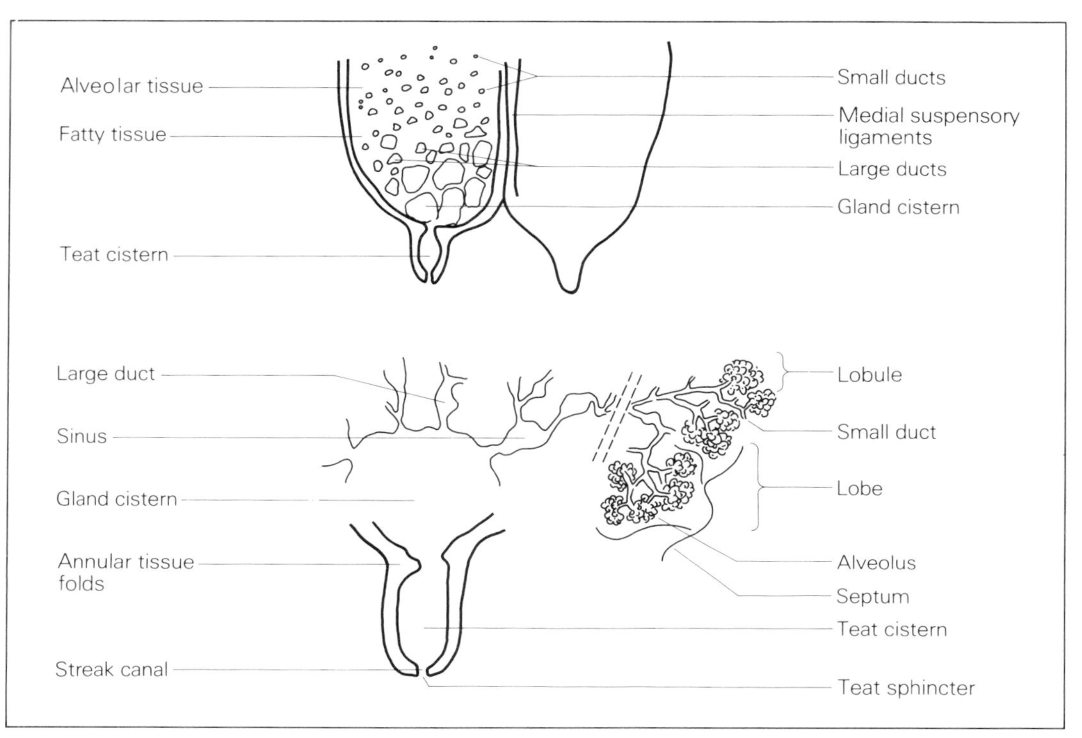


**Figure 2.5** Internal structure of the mamma (diagrammatic representation, alveoli not to scale)

The supply of blood and nerves to the mammae is depicted in Fig. 2.6(*b*) and (*c*). The blood supply passes through the inguinal canals; there are two in the cow, right and left. The mammary artery enters each half of the udder just forward of the rear teats and branches to form cranial and caudal branches serving the fore and rear mammae respectively. The arteries branch to form the capillary bed surrounding the alveolus, the prime function of the arterial system being to provide the milk synthesizing cells with a continuous supply of nutrients from which to synthesize milk. The veinous system drains from the capillary bed of the alveolus and mirrors the arterial system in forming cranial and caudal mammary veins which link to exit via the inguinal canal. Some blood enters and leaves the rear mamma via the perineal route, but this comprises a small proportion of the total flow. The most obvious feature of the blood vessels of the mammary glands of the cow are the 'milk veins', the subcutaneous abdominal mammary veins, which leave the gland under the skin but exterior to the abdominal wall. Their passage is tortuous and in some cows the veins are very prominent, entering the body cavity through a round orifice in the abdominal wall. This orifice may be readily identified by following the vein from the front of the udder and inserting a finger at the point of disappearance. It was once thought that the size of the milk veins was directly related to potential milk yield. It is not the case however, as blood supply is only one of the many elements controlling the rate of milk secretion; furthermore, the abdominal mammary veins drain only a small proportion of the udder.

The nervous system is not directly involved in milk secretion or milk removal, but is essential to the milking process which requires the active, conscious and willing participation of the central nervous system (brain) of the lactating female. The mammae themselves are supplied with nerves (Fig. 2.6*c*) which lead *to* the spinal column and brain *from* terminal endings in the glandular tissue and the skin and around the base of the nipple. These termini are particularly sensitive to touch, stretching and pressure, and convey to the central nervous system both that the mamma requires to be emptied and that the means of milk withdrawal (either machine or young) is present at the nipple. Nerve fibres bringing instructions *from* the spinal column *to* glandular tissues terminate at the muscles of the teat sphincter, some ducts and some blood vessels. At the time of milking, nervous impulses from the central nervous system induce muscle action which opens the sphincter to allow milk to be expelled from the gland and which also aids the propulsion of blood to the mamma and the transfer of milk down from the alveoli and through the ducts. It is a remarkable attribute of mammary glands, and a characteristic central to understanding the phenomenon of milk production, that the cells which secrete milk, the alveoli which are the functional elements in milk production, and the small ducts which transfer milk to the gland cistern, are not themselves in

**Figure 2.6** (*a*) Suspensory ligaments; (*b*) blood supply; (*c*) nerves of the functioning gland

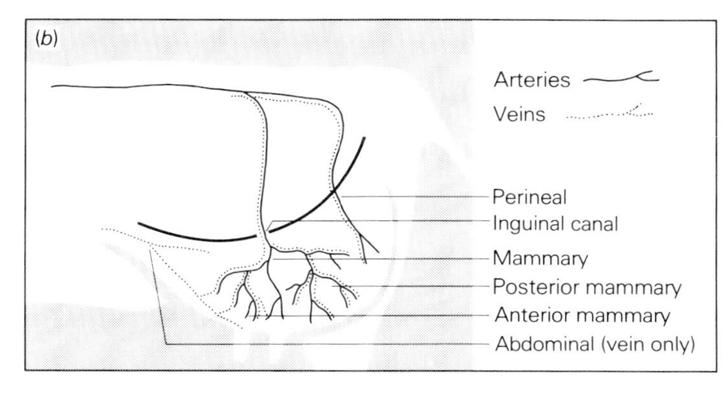


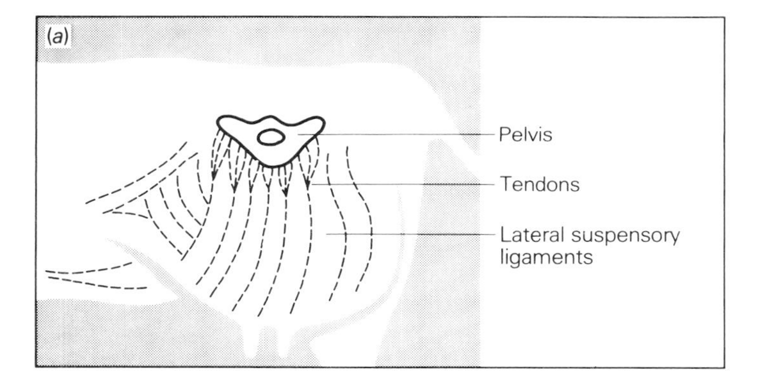


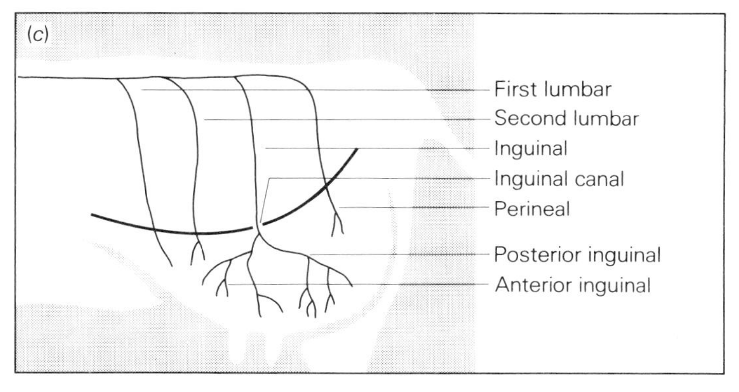

contact with any part of the nervous system. The activity of these elements of the mamma are controlled by hormones via the blood supply rather than being under the direction of the nervous system. Nevertheless, these hormones must be released into the blood stream as rapidly and completely as possible. The mechanism for hormone release is triggered via the nervous system at the level of both the brain and the nerves present at the skin surface of the mamma and in the region of the nipple.

At the microscopic level, the gland and teat cisterns and the large ducts are lined with epithelial cells. Passing from the inside – milk-side – to the outside, there is no secretory tissue but two layers of epithelium set upon a basement membrane; outside this lie longitudinal and circular muscle layers. As the ducts extend away from the gland cistern the amount of muscle diminishes. Toward the proximity of the alveolar lobules, muscle fibres are replaced in the small ducts by myoepithelial cells lying lengthwise along the ducts. These myoepithelial cells, in common with muscle cells, have the property of contraction. Myoepithelial cells are, however, unusual in that their contraction is initiated not by a nerve, as for muscle, but by hormone action. In addition to carrying myoepithelial cells as their contractile elements, the small ducts have but a single layer of epithelial cells placed upon the basement membrane. The single layer has some degree of milk secretion function, not possessed by the double epithelial layer. The closer to the alveolus, the more active the secretory

**Figure 2.7.** Structure of the alveolus showing: (*a*) blood capillary supply; (*b*) myoepithelial cells; and (*c*) lumen epithelium of secretory cells lying upon a basal membrane

cells of the small ducts.

The structure of the alveolus is shown diagrammatically in Fig. 2.7 as having an outer surface of delicate, loose tissue forming a capillary bed, below which myoepithelial cells are thickly gathered around the surface of the spherical alveolus. The size and shape of the alveolus has much to do with the extent of its filling with milk. Below the myoepithelium is the basement membrane upon which lies the single layer of secretory epithelium upon whose activity milk production depends. The function of the cells of the secretory epithelium is to draw constituents from the blood, synthesize milk and secrete the products into the hollow cavity (lumen) formed by the spherical nature of the alveolus.

## Secretion

There are three phases to secretion: synthesis by the cells of the alveolar epithelium, discharge from the epithelial cells into the alveolar lumen and storage in alveoli, ducts, sinuses and cisterns. Following milking, the lumina of the alveoli are empty, the epithelium is folded and the cells themselves largely devoid of any secretion. As the milk-manufacturing cells draw the raw materials for milk synthesis from the blood, synthesized secretory products fill the interior spaces of the cell and accumulate in the apical regions before leaving through the walls at the apex to fill the alveolar lumen with milk.

About 500 volumes of blood pass through the glands in the production of 1 volume of milk (this is about 10 tonnes of blood per day through the udder of a dairy cow). Some milk constituents move directly, without a change in form, from blood to milk; while others must be synthesized by the epithelium. Casein and lactose, for example, are found only in milk, and milk fat comprises glyceride combinations not found elsewhere in the body. Further changes in the composition of milk may take place by diffusion as the milk lies in the ducts. The approximate composition of cow's milk is given in Table 2.2. The major blood precursors of milk constituents are triglycerides, free fatty acids, acetate, glucose and amino acids. The energy-yielding glucose, fat and acetate

**Table 2.2** Percentage composition of fresh cow milk

| | |
|---|---|
| *Water* | 87.4 |
| *Total solids* | 12.6 |
| Fat | 3.7 |
| Solids not fat | 8.9 |
| Protein | 3.6 |
| Lactose | 4.6 |
| Ash | 0.7 |
| Ca | 0.13 |
| P | 0.10 |
| K | 0.14 |
| Cl | 0.11 |
| Mg | 0.01 |
| Na | 0.05 |

elements fulfil the dual role of raw material substrates for milk manufacture as well as that of fuel to drive the synthetic process (leaving the body as heat after the work is done). Most of the action is in the formation of glycerol from blood triglycerides and glucose, and in the formation of milk fatty acids from blood triglycerides, free fatty acids and acetate. Glycerol and fatty acids thus derived are joined together to form milk fat. Blood glucose, in addition to supplying energy and augmenting glycerol formation, is the precursor of the milk sugar, lactose. Blood amino acids are drawn together to make milk proteins, primarily casein. In energetic terms, the formation of milk fat, milk lactose and milk casein is relatively efficient (probably 75–85% efficient). In terms of the total economy of the functioning gland, however, the energetic costs of casein protein formation have the appearance of being rather higher and the efficiency rather lower (50–60%) on account of the rapid rate of turnover of the cells of the secretory epithelium, which are themselves largely protein in nature (Fig. 2.8).

Synthesized fat droplets accumulate in the cell that manufactures them, together with the water, protein, lactose, mineral and vitamin elements which make up the other milk constituents. Milk is secreted from the cell into the alveolar lumen probably in two ways. Apocrine secretion occurs where the cell contents are pinched off as the milk constituent is squeezed from the cell. The cell membrane generally remains intact, but some milk fat globules may be observed to have cell cytoplasm attached to them, suggesting some cell break-up as the fat droplet passes through. Eccrine secretion occurs where milk constituents pass through the apical cell membrane without any loss of cytoplasm; this is assumed to be the more usual method for passage of the non-fat milk constituents. In the normal course of events the final loss of ageing cells is likely to be associated with damage incurred in the secretory process of discharging milk from the cell into the storage spaces of the gland. A level of discarded cells as high as 0.25 million per ml of milk is usual in normal milk, while the rate of cell loss may increase to 1 million per ml in cases of udder damage or mastitis.

Milk is synthesized at the rate of about 1–2 ml/g glandular tissue per day. Pressure within the mamma is low following milking and for the next hour or so, but increases from then on. Milk is discharged from the cell at an inverse rate to mammary pressure; milk synthesis is therefore most rapid in the first few hours after milking and slows down progressively with time from the last milking. The time to be allowed between milkings is a decisive element of dairy management, manifesting itself in the choice both of number of milkings completed in a 24-hour day and of milking intervals. As mammary pressure is the cause of reduced secretion rates, the matter turns on the point at which accumulation of milk in the udder is such as to create the degree of pressure necessary to effect a significant drop in the milk synthesis rate. This pressure level may be as high as 50 mmHg (6.6 kPa), and the

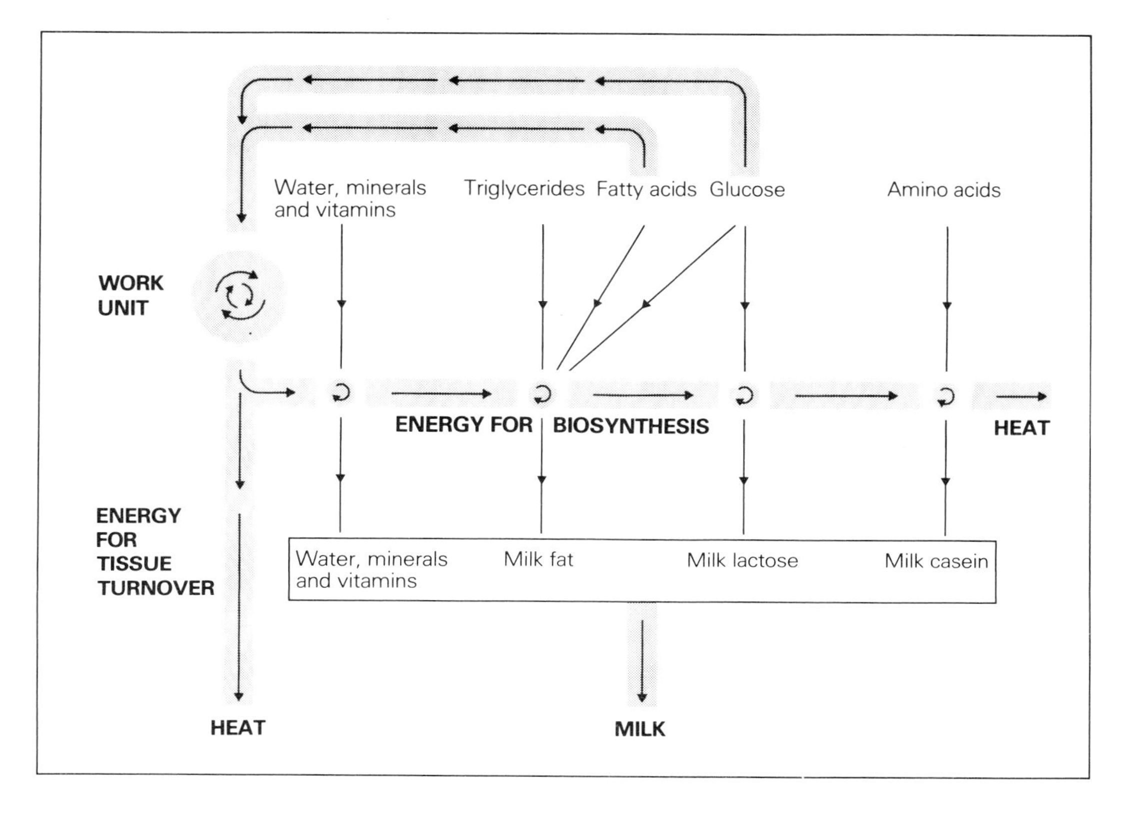


**Figure 2.8** Milk synthesis

time interval from the last milking at which it is attained will relate closely to the potential milk yield of the cow in comparison to the storage capacity of the mammae. The higher the yield and the lower the storage capacity the more frequently must milk be removed if reduction in secretion rate is to be avoided.

## Hormones involved in lactation

Like growth in general, mammary growth is under the control of the growth hormones produced by the anterior pituitary gland. Post-pubertal mammary growth is controlled in addition by the ovarian hormones, oestrogen and progesterone, connected with the three-week reproductive cycle. Oestrogen is thought to be required particularly for duct growth. The two hormones together, with the addition of prolactin (also from the anterior pituitary), are needed for the growth of the alveolar lobules that occurs in pregnancy. The milk secretory activity of the epithelial cells of the alveoli is especially stimulated by the action of the hormone prolactin – so named because of this function.

The manufacture of milk and its subsequent secretion occurs before the young is born. Alveolar lumina fill during most of the second half of pregnancy, in the last 20 days the secretion rate being particularly rapid. Milk is, however, not normally removed from the gland until the young is born and sucks. Some high-yielding dairy cows can suffer a considerable degree of discomfort in the final week of pregnancy and in these cases 'pre-partum' milking is practised in some dairy herds for particularly susceptible cows. It appears that the presence of milk constituents in the alveoli prevents full-scale milk synthesis and holds up lactation. The hormones oestrogen and progesterone, from the ovary and placenta, whilst *encouraging* mammary growth, also *discourage* lactation; but the level of discouragement is not particularly great in the dairy cow as evidenced by the phenomenon of pre-partum milk production. In addition, the high ratio of progesterone to oestrogen in pregnancy inhibits the hormone prolactin; as prolactin stimulates milk synthesis, pregnancy thereby inhibits lactation. This factor is particularly important for species which, like the dairy cow, need to be pregnant and lactating simultaneously. Progesterone also holds up lactose synthesis; as lactose is the main osmotic factor in milk, the presence of lactose in the alveolus is the trigger which initiates the secretion of water and the other water-soluble milk components into the lumen. At the end of pregnancy (around 30 hours before the birth) oestrogen levels rise and progesterone levels fall. The hormone oxytocin is secreted from the posterior pituitary, causing uterine muscle contractions, and eventual expulsion of the young. These changes around the time of the birth remove the blocks to prolactin and milk synthesis, and the lactation can begin.

Mammalian lactation is only relevant in the context of a

beneficiary; so the final initiator of lactation must be the actual removal of milk. Milk removal starts off lactation in a number of ways: by removing back pressure to the passage of synthesized milk from the cells into the alveolar spaces; by removing the milk constituents whose presence in the cell and alveolus retard the synthesis of more milk products; by releasing oxytocin which indirectly encourages the hormones favourable to lactation; and by stimulating into action the hormonal complex which is involved in accelerating the metabolic activity of the gland. In short, the change in hormone balance at birth places hormones positive to lactation in the ascendancy and the removal of milk from the gland stimulates the secretory epithelium to synthesize more. The cells of the epithelium proliferate particularly rapidly around the time of the birth and the synthesizing activity of each cell is also enhanced. The increase in milk yield which occurs in the early stages of each lactation is thus associated with both increasing cell number and increasing cell activity. Total lactation yield results from the combined effects of the amount of milk produced daily and the duration of the lactation. Increasing lactation yield after the birth is mainly a function of the level and efficiency of the cow's metabolism. The metabolic activity depends in its turn upon the level of hormones which encourage the biochemical reactions needed for milk synthesis, and which promote the supply of nutrients to the manufacturing cells in the gland. Given an unlimited supply of the raw materials for milk synthesis in the blood, the rate of increase in yield at the beginning of the lactation is next a function of the rate of withdrawal of milk from the gland; that is, the efficiency of the milking machine equipment, or the number, size and appetite of the sucking young.

Under the positive influence of hormones and the stimulus of milk removal, milk yield rises to a peak some time between the fourth and twelfth week of the lactation in the machine-milked dairy cow. For the rest of the lactation the yield is in continual decline. Clearly, the higher the peak and the slower the rate of decline, the greater will be the lactation yield. The height of the peak is the most important factor influencing the level of milk production from dairy cows, and the rate of decline as the lactation progresses is the second most important factor. The hormone prolactin helps to maintain the functional activity and the structural integrity of mammae, while the other hormones involved in nutrient supply, epithelial replacement and milk synthesis are clearly prerequisite to continued milk production.

Just as there would be no lactation in the absence of any means of taking milk from the gland, then, conversely, sucking or machine milking is a potent force for maintaining the level of the daily milk yield. Milk removal from the alveoli triggers the secretory epithelium to synthesize more milk; just as a build-up of milk in the alveoli reduces milk synthesis by metabolic inhibition and by back-pressure effects. Efficient milk removal at sufficiently short time intervals in relation to the synthesis rate and storage capacity are therefore central to maintaining the yield of a dairy cow at the highest possible level. Again, the collapsing of the alveoli such as occurs when milk is taken from the mammae helps to maintain their integrity and aids the circulatory system. Milk removal has further positive effects, via the nervous system, upon the balance of hormones encouraging milk production, and there is also a sharp rise in prolactin during milk removal. Prolactin release is probably enhanced by oxytocin, that fascinating hormone found transiently in the bloodstream while milk is flowing from the mammary gland.

Although the lactation must inevitably decline and dwindle away (in 12–20 months in the case of the dairy cow), one of the major factors increasing the decline in milk production as the lactation progresses is the imposition of another pregnancy. In commercial dairy herds this usually occurs between 8 and 16 weeks after the lactation has begun. By about the fifth month of the pregnancy the daily rate of reduction in milk yield begins to sharpen perceptibly. Pregnancy is antagonistic to lactation in three different ways. First, the pregnancy hormone progesterone discourages the lactation hormone prolactin. Next, the foetus and the associated tissues of the uterus and placenta compete with the mammae for those hormones which stimulate milk synthesis. Last, there is an ever increasing demand by the foetus and uterus for nutrients, which reduces the supply going to the mammary gland. It is in the nature of things that the foetus, as the new generation, will have priority over the previous youngster who should now be competent to cope with food sources independent of its mother.

The ability of the dairy cow to become pregnant while lactating is tacitly accepted, and indeed is a necessary adjunct to the need to breed annually together with supporting a 280-day pregnancy. Such a facility is, however, by no means universal. The sow, for example, shows lactational anoestrus and cannot readily conceive while she is lactating. The human female, with characteristic unpredictability, can show variable anoestrus; thus, while it is usually the case that breast feeding also acts as an efficient contraception, this is not invariably so. Although the fuller the lactation the stronger is likely to be the anoestral effect.

Milk production decline to final cessation is as a direct result of a reduction in the number of secretory cells (epithelial cell replacement rate failing to keep up with the rate of cell loss) together with a reduction in the synthetic activity of the individual cells which remain. As lactation proceeds towards its conclusion, the sucking stimulus weakens and the demands of pregnancy become greater. There is a reduction in those hormones beneficial to lactation maintenance and a contrary increase in those prejudicial to lactation; the impending birth thus heralds the end of the current lactation as well as the beginning of the next.

In the absence of a pregnancy the final end to lactation occurs as a result of the general senility of gland tissue which results from continued loss of secretory cells without

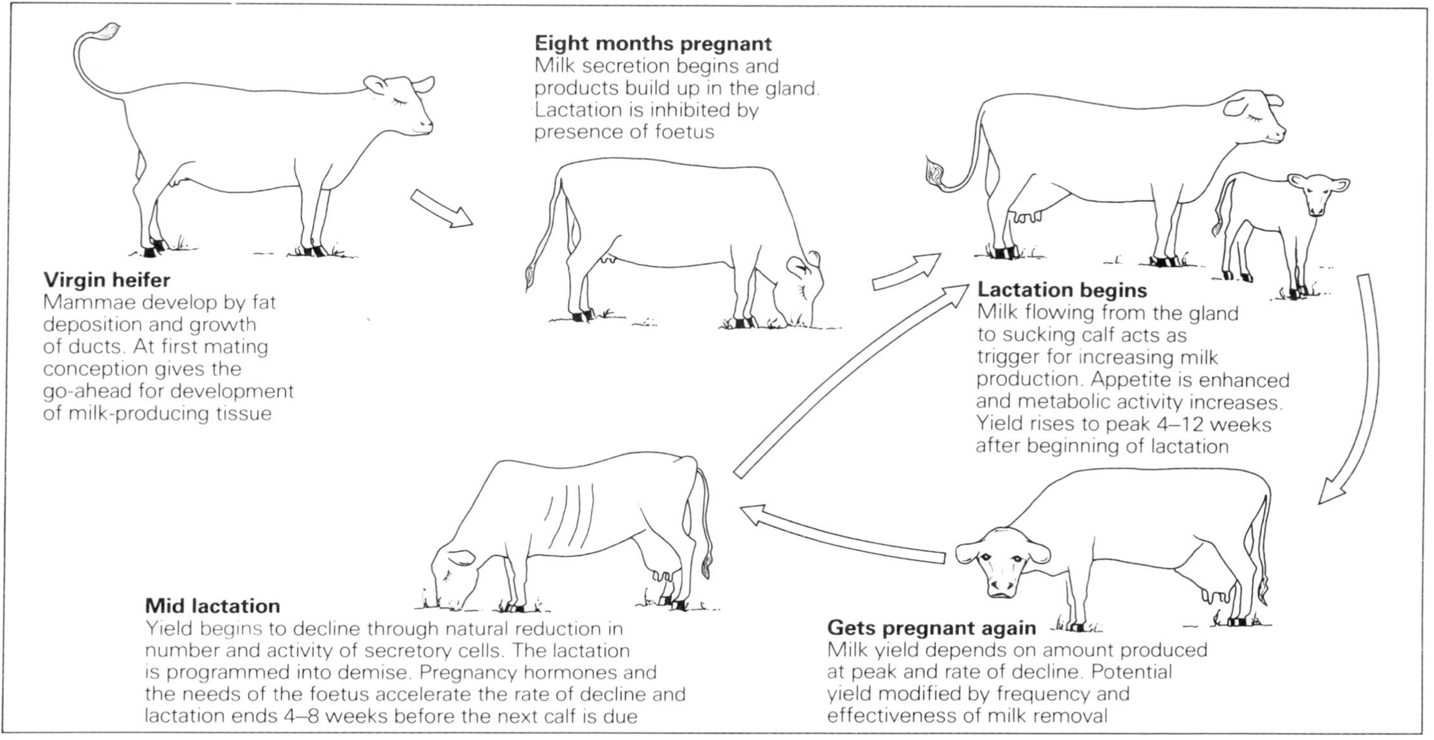


**Figure 2.9.** The rise and fall of a lactation

replacement. However, the lactation can be ended before this time by failure to withdraw milk, or its decline and fall brought forward due to inefficient machine milking or loss of sucking stimulus caused by the absence or debility of the youngster. Secretion will be inhibited by the presence of the end products of milk synthesis in the alveoli and a build up of back pressure, with the consequent termination of the lactation. The final act of the secretory epithelium is to reabsorb milk constituents synthesized but not removed. The epithelium then regresses, leaving the alveolar lobule merely as a deflated structure devoid of active secretory tissue, but maintaining a few inactive epithelial cells and all the other elements as already shown in Fig. 2.5. In order that the next lactation should not suffer from the dilapidations to secretory tissue caused by the previous lactation, the gland requires a quiescent period during which the degenerate epithelium is rejuvenated, first under the influence of the hormones stimulating mammary growth and next under the influence of the hormones positive to milk secretion. A summary of the factors involved in the rise and the demise of a lactation is illustrated in Fig. 2.9.

## Milk ejection

Milk is taken from dairy cows by machine usually twice daily; calves suckle naturally six to ten times daily. The release of milk from the mamma to the machine is itself under hormonal control. This is rather surprising inasmuch as the propulsion of milk from the mammary gland would come under the definition of an acute reaction, taking some 10–20 seconds for complete gland emptying in the case of the sow and 3–5 minutes in the case of the dairy cow. Such immediacy is usually associated with nervous reactions whereas hormonal reactions are often chronic in nature; thus the flexing of an arm muscle under nervous control may be contrasted with the growth of that muscle which is under hormonal control. Hormones with such singular abilities of evoking acute reactions rather than chronic ones are clearly of great interest in behavioural and physiological terms. What appears to happen is that the stimulus of sucking or milking results in a nervous reflex from the mammary gland to the brain, which brings about the release of the hormone oxytocin from the posterior pituitary gland. Oxytocin causes the contraction of myoepithelial cells. Myoepithelial contraction brings about implosion of the alveoli and milk is forcefully ejected and rushes down into the ducts. Pressure build-up from simultaneous contraction of all of the alveoli in the mamma causes milk to flow rapidly out from the milk storage compartments of the gland when the teat sphincter opens. Myoepithelial cells may also contract in response to physical stimuli. Butting, nosing, pushing and squeezing of the mammary gland not only helps the flow of milk out of pockets and sinuses, but also brings about a limited milk ejection

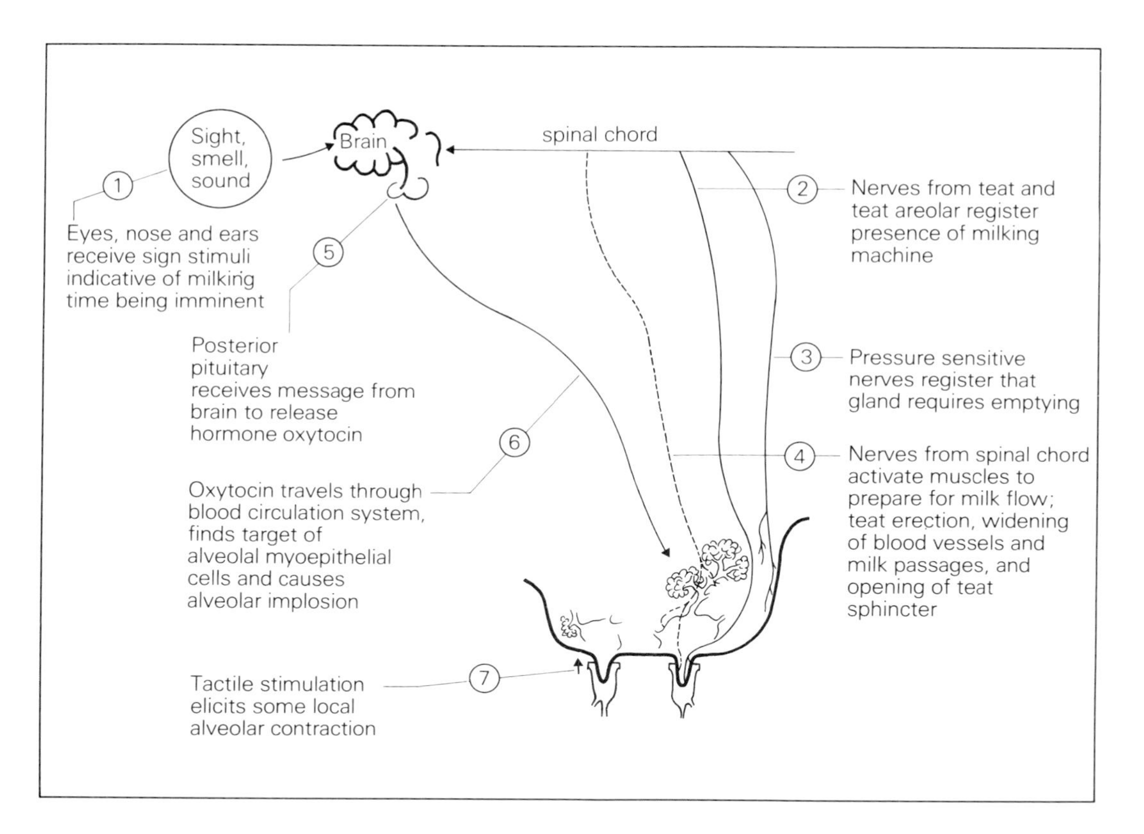


**Figure 2.10.** The elements involved in the milk ejection reflex

response from local alveolar lobules.

The 'neurohormonal' milk ejection reflex is described pictorially in Fig. 2.10. The action of the mouth of the young on the teat and around its base excites the pathway for nervous response. The impulses pass via the spinal cord to the brain, in particular to the paraventricular nucleus (PVN) of the hypothalamus. The PVN is concerned with oxytocin production and release. As a result of the nervous impulse acting on the PVN, oxytocin is released into the bloodstream from its place of storage in the posterior pituitary. Other stimuli, not tactile, may also elicit the release of oxytocin. These stimuli come through the ears, eyes and nose and are connected with imminent milk removal. For example, the presence of the offspring, or the paraphernalia of the milking parlour in the case of the dairy cow. This neurohormonal reflex can be disrupted by disturbance. Adrenalin may bring about constriction of the blood vessels, reduce the blood flow in the mammary gland, and thereby prevent oxytocin reaching the target site of the myoepithelial cells. Most important of all is the ability of the brain to prevent oxytocin release from the posterior pituitary by blocking the impulse if suitable conditions for milk removal do not prevail. In addition to stress factors, unsuitable conditions for milk removal might be an insufficient time lapse between suckings, faulty milking equipment, an unfamiliar milking machine routine, the absence of a familiar element in the routine (such as feeding or udder washing), or even an unfamiliar operative in the parlour. The disturbance of milk ejection by relatively trivial agents bears evidence to the nature of the process initiating milk flow from the mammae, either by natural suckling or by artificial machine milking, being a two-sided and voluntary affair; both the young or the machine, *and* the mother must be *simultaneously* ready, willing and able.

# Natural milking

## 3

Mammalian young are suckled at relatively even intervals through the day, the time elapsing between feeds depending upon species and upon the age of young (Figs 3.1 and 3.2). Calves suck naturally some 6–10 times daily (the number of sucklings decreasing with age), while piglets suck 20–24 times. During the interval between milkings, milk secreted from the epithelium into the lumina of the alveoli is fed into the small ducts to be stored by large ducts, sinuses and the cistern. In the small ducts and alveoli there is, with the exception of the immediate post-milking phase, a fairly constant amount of milk; but the amount in the cistern will depend upon the amount of time that has lapsed since the last milking. With the dairy cow milked twice daily, 30–50 per cent of the total yield is contained in the gland cistern and its closely associated sinuses, the higher proportion following the longer milking interval. As milk accumulates, pressure in the mammary glands rises. The level of pressure, and thereby the frequency at which milk should be removed, will depend upon the degree of cisternal filling. The sow has a proportionately much smaller gland cistern than the cow, and it is significant that piglets suck much more frequently than calves. The *Bos indicus* (Zebu) cattle types are also characterized by having a small cisternal capacity as compared to *Bos taurus* (European) cattle.

To help expulsion of milk from the mammary gland there are two types of contractile tissue: myoepithelial cells and smooth muscle. Myoepithelial cells are found around the

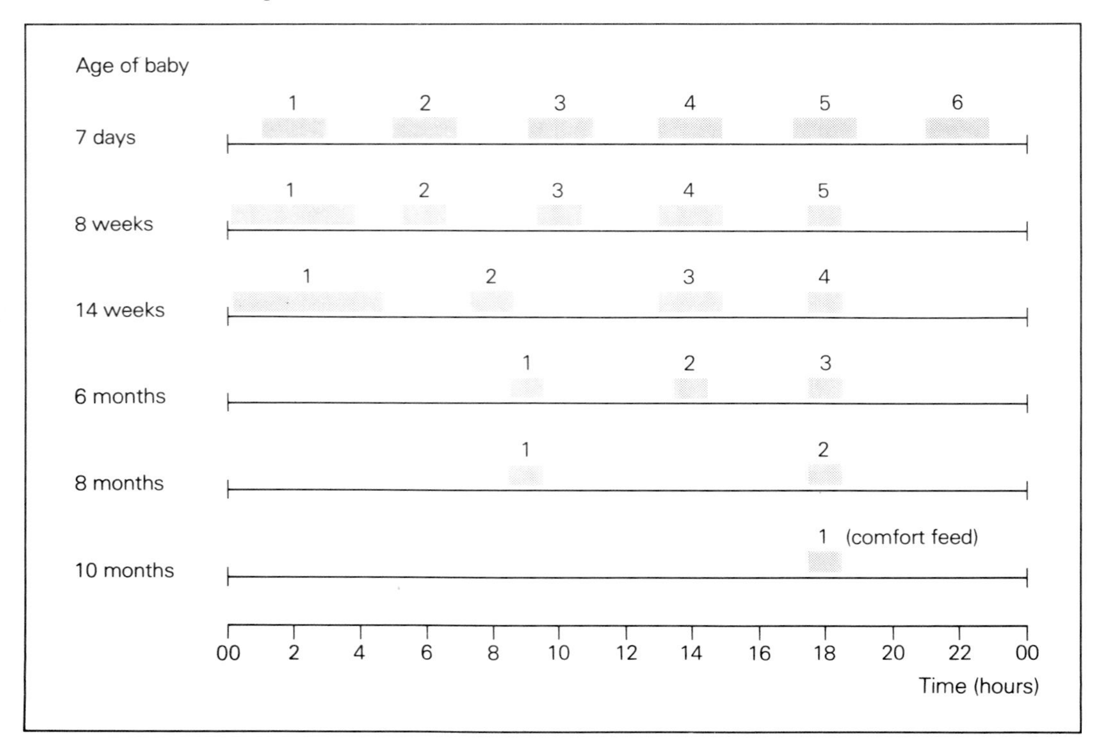


**Figure 3.1** Examples of frequency of breast feeds given to a demand fed human baby (E.C.W.). ▒ indicates the time period within which a single feed usually occurred. For example, at 14 weeks the first feed of the day was sometime between midnight and 5 am

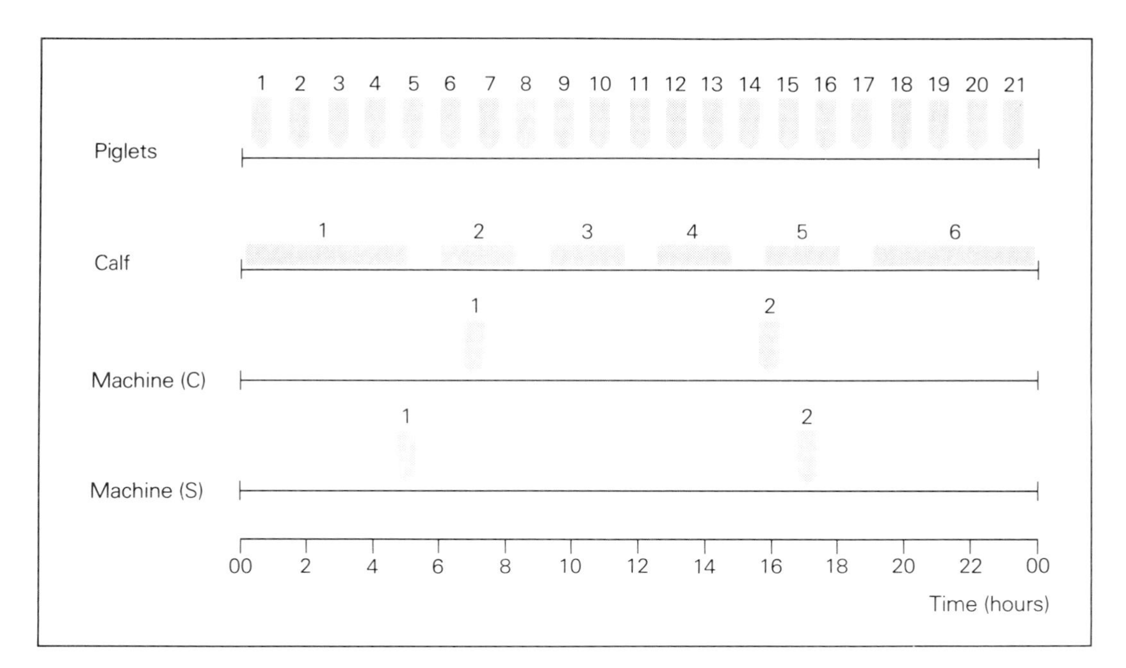


**Figure 3.2** Frequency of milkings by piglets, calf and milking machine. Machine (C) shows the frequency forced by working a conventional farm day (7.00 hrs–17.00 hrs). Where a shift work system is used (Machine (S)), the intervals allowed between milkings can be more equal, which removes the problem of a short (day) interval and a long (night) interval

alveoli dispersed in a stellate arrangement to effect a squeezing action, and along small ducts in a longitudinal arrangement to effect a shortening and widening action (Fig. 2.10); myoepithelial cells do not tire easily and will respond to recurrent excitation. That the same form of myoepithelial cell should, through its disposition, effect two quite opposite reactions is a remarkable illustration of biological economy. When in a stellate arrangement and placed around the spherical alveolus, contraction of the limbs of the myoepithelial cell will bring about a rapid *implosion* of the cell, collapsing it to a much reduced size. While arranged longitudinally along the small ducts, contraction of the myoepithelial cell brings about an *expansion* of the duct by shortening and widening. This aids the expulsion of milk from the collapsed alveolus by simultaneously increasing the size of the passage through which the milk must pass. In contrast to the myoepithelial cells, smooth muscle is found around the blood vessels, around small and large ducts and around the sinuses (mostly in small amounts) and in the form of annular muscle sphincter arrangements particularly in the teat.

The smooth muscle is excited by nerves from the spinal cord which pass down through the inguinal canal, spread through the skin of the mammary glands and the teats, and terminate in the smooth muscle of ducts, teats, sphincters and blood vessels. Myoepithelial cells have no nerve connections, but are excited almost entirely by hormone action (apart, that is, from some degree of activation by physical means – pressure from manipulation of mammary tissue may result in contraction of myoepithelial cells in the area of contact). The hormone oxytocin released in a burst from the posterior pituitary brings about the dramatic contraction of myoepithelium about 20 seconds after release of the hormone from the pituitary. A time lapse of approximately this length is common to the cow, sow, ewe and human female. Oxytocin has a circulating half-life of about 1–2 minutes in the cow and its action lasts for 2–5 minutes after secretion. Equivalent values for sow and ewe are probably less than 1 minute, while in the cat, exceptionally, oxytocic activity may last for 5–10 minutes. Activity time is dependent on the amount of oxytocin released from the pituitary at each burst, and the number of bursts within any one period of time. Thus in terms of contractive efficacy, one burst of oxytocin lasting for 2 minutes is roughly equivalent to two bursts lasting for 1 minute each.

During normal milking sessions in the cow and pig, oxytocin is usually released as a single dose, although a double dose is not infrequent. Multiple doses appear more prevalent in ewes and it is sometimes the practice to milk sheep twice at an interval of 10 minutes or so. Oxytocin release by a series of drips appears to occur in the immediate period after birth when the newborn young are not sufficiently strong to suckle at regular intervals, but rather require the supply of milk to be continuously available from the mammae. It is conjecture as to which is more effective: the release of oxytocin in one large dose, or in a multiple series of smaller doses. If milk is not all removed within the active circulation time of a single release, then as myoepithelial cells will go on responding to repeated doses of oxytocin, a number of releases may be as efficient as a single one. For dairy cows milked to a parlour routine, one requires not just effective milk removal, but also rapid milk removal. So a single release of a large dose of oxytocin is probably most expeditious. This should sustain myoepithelial contraction for whatever period of time is necessary for complete gland emptying.

The short half-life of oxytocin is consequent upon its removal from the blood stream by rapid catabolism in the mamma itself, and in the kidney, liver and uterus. In addition to its rôle in milk ejection, oxytocin is active in other situations in the mammalian body involving convulsive muscle contraction; for example, of the uterus during parturition and probably coitus. Milk may occasionally be seen to drip from mammae at the moment of coitus. Conversely, uterine contractions may occur simultaneously with milk release. As milk withdrawal sessions are readily entered into by mammals, it is sensible to assume that the suckling female derives some positive and acceptable stimulation from the experience. The positive nature of the experience may be associated both with the reversal of the unpleasant sensations connected with the build up of pressure and tenderness in the gland as it fills with milk, and also with the pleasant sensations of the effect of oxytocin. In retrospect, it appears only reasonable that if the brain of the animal is to be positively involved in the activity of milk flow, then milk flow should provide positive feedback to the brain. This element of conscious involvement and willing participation of the female is absolutely essential to all successful milking routines. Nevertheless, the degree and quality of the positive sensations associated with milk release appear to differ widely between individuals, and within a single individual, depending upon circumstances.

An effective parlour routine for a dairy cow depends to a considerable extent upon the sympathetic handling of the animal and an understanding of her physiological needs before, during and after the process of milk release.

## Initiation

Nerves run from endings situated around the teat (which are touch sensitive), at the gland surface (sensing tension, temperature and pressure) and in the gland tissue (registering intramammary pressure). Impulses from these nerve endings join the spinal column from where reflex responses may, within 5 seconds, cause teat erection and squeezing of ducts and cisternal walls. In this way, smooth muscle action moves milk from the ducts to the sinuses, controls the flow of milk into the cistern, and at the same time prepares for the free flow of milk from alveoli when the myoepithelium contracts. Another, independent, nervous reflex controls relaxation of the sphincter at the teat orifice to allow milk flow to the outside. Other pulses pass from the tissue of the mamma along the nerve fibres to the hypothalamus in the brain where the need for milking is registered and the presence of the means to do it is noted. This results in behavioural responses allowing sucking by the young, or attachment of the milking machine. Information then passes from the hypothalamus to the posterior pituitary which results in the release of oxytocin.

Nerves running from ears, nose and eyes may relay information to the brain about the imminent presence of what the animal has come to learn are the means of achieving milk removal. By such conditioned stimuli, the nervous link between the mamma and the brain is short circuited; the resultant release of oxytocin is, however, the same. Nerves also relay information that is disadvantageous. For example, milk release may be held up by pain in the mamma, or by the presence of factors prejudicial to milk removal – such as unaccustomed noises, predators, strangers, a change in routine or unsympathetic handling. Relay to the brain of such information will result in nerve pulses passing from the brain to mammary smooth muscle, causing constriction of the blood vessels, closure of the small ducts and constriction of the milk passages. A reduction in the size of the blood vessels prevents blood-borne oxytocin reaching the myoepithelium, while closure of small ducts and milk passages results in a hold up of milk in the glands. Secretion of the hormone adrenaline would also cause constriction of the blood vessels. The most significant effects occur at the brain itself which, when conditions are unsuitable, simply prevents oxytocin release from the posterior pituitary and so terminates at source any chance of milk ejection.

## Suckling

### Passive withdrawal

This phase of milking is so called because the *active* participation of the mother is not required. The young approaches the dam and begins to massage the mamma. Both the presence of the young and pressure in the gland are needed simultaneously to trigger behavioural responses conducive to suckling. The mother–young coupling may be initiated by either party, depending upon the relative urgencies of gland discomfort (mother) or hunger (young). The mother becomes progressively less tolerant of the attentions of her offspring as he gets older. When milking begins, the teat sphincter at the tip of the nipple opens and milk may drip out. Although the teat is taken into the mouth of the sucking young by means of negative suction pressure, this is primarily to maintain the position of the teat in the mouth and not to withdraw milk. The tongue is usually partially wrapped around the teat and extends from the base to the tip, even including some of the mamma itself. The tongue makes a rippling, stroking, movement such as to drag milk down from the base (top) of the teat to the teat sphincter at the tip (bottom). This action should be compared with hand-milking where pressure is also from above and positive, whereas in the case of machine milking the pressure is from below and negative. In the case of the pig and also the human, where the gland cistern is small (holding only about 10 per cent of the total milk), little milk is obtained during this phase. In the dairy cow the cistern is large and 40–50 per cent of total milk may be obtained during the passive withdrawal phase of milk removal. Continued butting and sucking by the youngster may also stimulate by mechanical action contractions of a few of the alveoli and small ducts. Whilst this passive withdrawal of cisternal milk is taking place, nerve pulses pass from the teat in response to the stimulation of sucking, together with impulses picked up from ears, nose and eyes. These pulses enter the hypothalamus which then passes the information on to the posterior pituitary and oxytocin is released into the blood stream. Meanwhile, some 5 seconds after stimulation, the nervous reflex has brought about duct squeezing by smooth muscle. The milk in the ducts is pushed into the gland sinuses and cistern. Oxytocin now reaches the alveoli and ducts, which implode and force the milk under considerable positive pressure down into the lower regions of the gland and out through the teat orifice. Hence the term ‘milk ejection’. Patterns of milk flow from mammae of the sow and the cow are illustrated in Figs 3.3 and 3.4.

## Milk ejection

The milk ejection reflex takes place 20–40 seconds after the initial stimulation of the mamma. Upon contraction of the myoepithelial cells, the alveoli collapse and milk is forcefully ejected into the small ducts. The ducts shorten and widen and the milk rushes through into the sinuses and gland cistern. Only alveolar and ductal milk is expelled by the action of oxytocin on the myoepithelium; there is no contraction of large ducts or cistern. Intramammary pressure rises due to milk being piled into the cistern. The sphincter at the end of the teat is opened and milk flows rapidly out from the gland. In the cow, the sphincter is relatively strong, as indeed it needs to be to hold back the weight of milk stored in the gland cistern for the period between milkings. In the sow and the human female, a small cisternal capacity removes the need for a strong sphincter; most of the milk in the gland is held in the alveoli and ducts and only likely to leave the gland following collapse of the alveoli when the young are present at the teat. In the case of the cow, then, the strength of the sphincter and the control of its muscles by nervous impulses figure largely in the mechanics involved in drawing milk from the mamma. Both the calf and the milking machine must not only ensure that the sphincter opens effectively to allow milk flow, but they must also take account of the fact that, like all muscles, those of the sphincter must rest. When the muscles are resting, the sphincter closes. The teat sphincter appears to go through an opening and closing cycle about once every second.

During ejection, milk flows freely into the area of lower pressure in the mouth of the young. The function of the mouth during this phase is not to suck milk from the gland but rather to facilitate the active flow of milk from the gland which occurs under pressure created by milk forced down into the

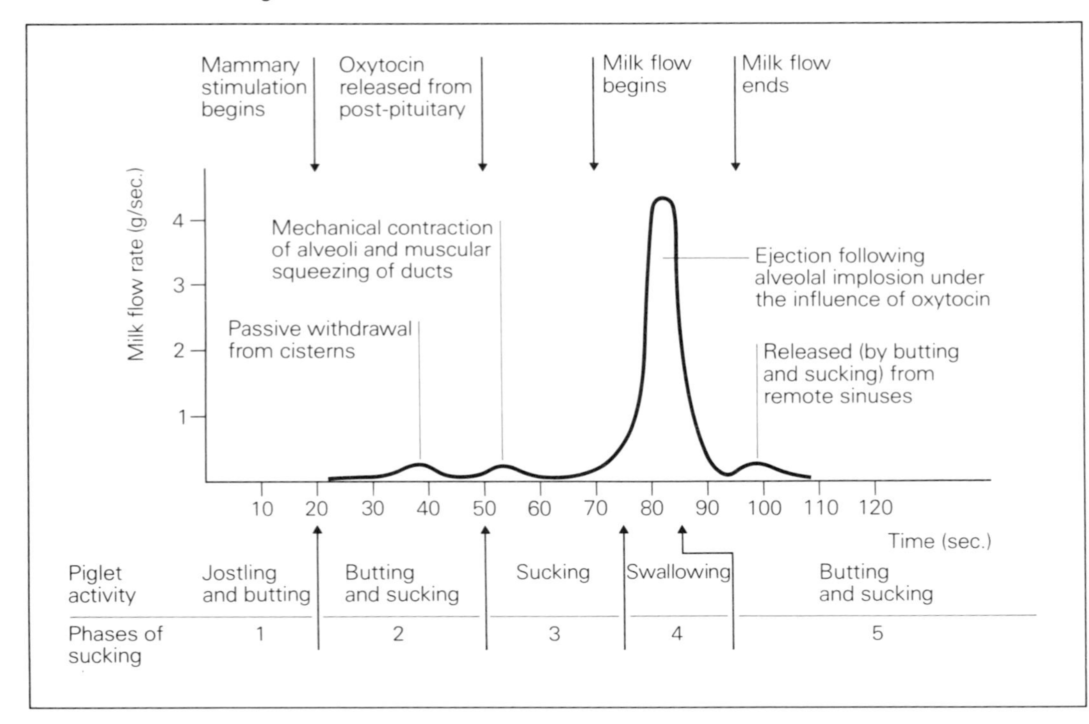


**Figure 3.3** Pattern of milk flow from mamma of sow

Maximum rate of milk flow from mamma: 4 g/sec.
Maximum rate of milk flow from sow: 40 g/sec.
Milk yield/mamma/suckling: 40 g
Milk yield/suckling: 400 g
Sucklings/day: 20
Milk yield/day: 8 kg
Milk yield/kg sow body: 50 g

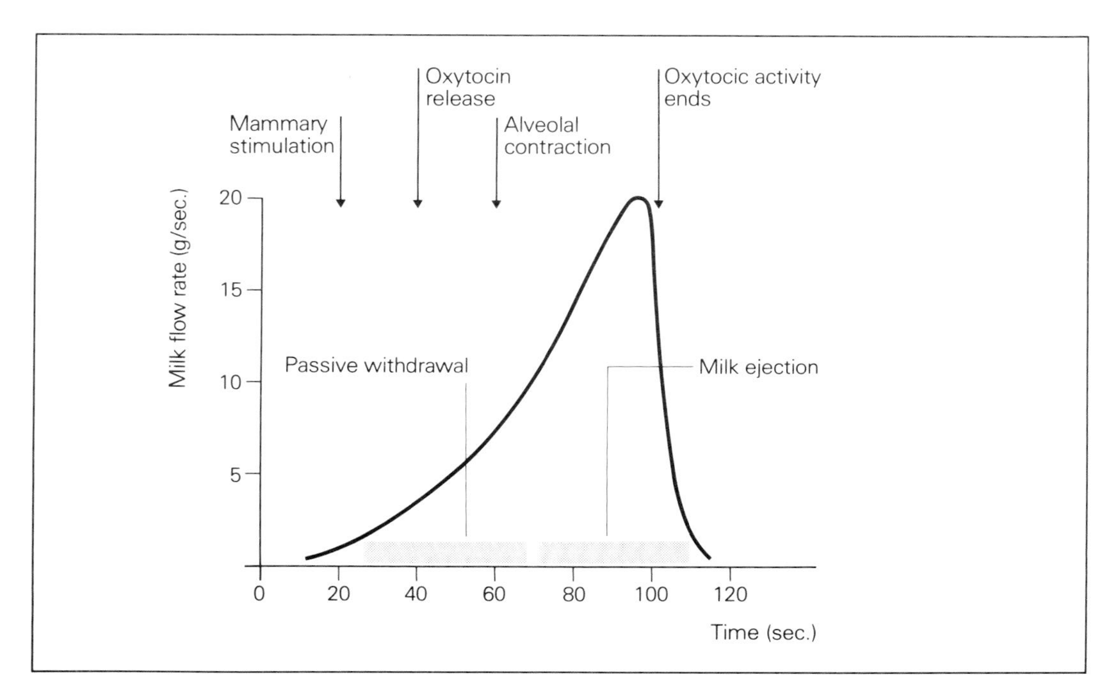


**Figure 3.4** Pattern of milk flow from mamma of cow

Maximum rate of milk flow from mamma: 20g/sec.
Maximum rate of milk flow from cow: 80 g/sec.
Milk yield/mamma/suckling: 500 g
Milk yield/suckling: 2000 g
Sucklings/day: 8
Milk yield/day: 16 kg
Milk yield/kg cow body: 30 g

cistern from imploded alveoli. Calves, piglets and human babies may be seen to swallow rather than suck; while the tongue acts as a chaser for the milk, not as a means of its withdrawal. The effective contraction time of myoepithelial cells and the consequent alveolar collapse time is 2–5 minutes in the cow, 10–30 seconds in the sow and perhaps 2–10 minutes in the human. After ejection, the alveoli expand back into shape and milk flow ceases. If all the milk is not yet withdrawn from the gland, reverse flow occurs from the sinuses into the ducts and from the ducts into the alveoli. Milk flow may only be re-initiated by another release of oxytocin. It is quite possible for more than one burst of oxytocin during a suckling; indeed it must necessarily be so where a single offspring is to draw milk from two or more mammae and the time taken to empty just one of the mammae is more than half of the total alveolar collapse time. It may, for example, take a calf 1–2 minutes to suck out *each* of his mother's four mammae. Although the action of blood-borne oxytocin upon the myoepithelium must be general, affecting all mammae simultaneously, the sucking young of the cow or the human must deal with the mammae consecutively. The general nature of oxytocic action has its implications for animals with multiple young, as all the young must be present at the udder at the same time. With pigs, the first phase of suckling may be a protracted affair, the mother being unlikely to allow milk release before the whole litter is settled down each to his own teat, ready to receive. In the case of the human female it is noticeable that the sucking infant at one breast can elicit a quite copious flow of milk from the other. A single large dose of oxytocin has been demonstrated in the blood of machine-milked cows and suckled sows just prior to milk ejection. Other secondary doses, usually only one other but occasionally two others, can also occur, but by no means do so invariably. The ability of the mammal to release oxytocin from the posterior pituitary is strictly limited; perhaps because the trigger mechanism becomes jaded, or because the store of hormone is depleted. In any event, a refractory period follows each session of suckling or milking, during which time milk ejection by oxytocic activity is precluded. If milk is not removed during the time when oxytocin is active, it will not again be available until the system is adequately refurbished. There are many similarities, in these respects, between milk ejection and coitus; this is understandable as oxytocin appears to be an active participant in both. The phenomenon of a refractory period after milk release is of great significance in the planning of a parlour routine. It is vital to capitalize rapidly and effectively upon a successfully stimulated cow, and to remove all the milk when it has been made available; there appears to be limited possibility of a second chance.

Alveolar and ductal milk removed following the milk ejection reflex comprises the milk which could not be withdrawn passively nor ejected by mechanical stimulation of the myoepithelium nor squeezed from the ducts by muscular contractions under nervous control. The importance of milk ejection to milk removal therefore varies between species, particularly according to the proportion of milk held in the storage elements of the glands and therefore available by passive withdrawal alone.

In the European dairy cow the milk ejection reflex is probably essential for only 50–60 per cent of the total yield.

For the Zebu or buffalo, or indeed the pig, milk ejection is vital to effective milk removal as only 10–20 per cent of total milk is held in the gland cisterns of these species. This characteristic can mitigate against the use of mammals with small milk storage capacities as commercial dairy animals, particularly if there is also any unwillingness to participate freely in the release of milk for purposes other than feeding their own offspring.

### Termination

At the end of milk removal the alveoli relax (expand) and the sphincters close. The suckler will attempt to draw the last drops of milk from the gland by returning to massage, butting and vigorous sucking to try to bring about a further milk ejection reflex (oxytocic, mechanical or nervous) and to spill out any milk which may have been held up in inaccessible pockets and in the more remote sinuses of the mammary tissue. Between 5 and 10 per cent of total milk synthesized will unavoidably remain in the gland, milk removal never being completely effective. This residual milk is not any loss to production, as it joins the new milk synthesized for the next milking session. Ineffective gland-emptying resulting in more than the minimum of residual milk does, however, have serious repercussions if the extra milk causes pressure build-up before the next milking sufficient to reduce the synthesis rate. This latter is most likely if milking times are rigidly fixed or where the refractory period for milk ejection prevents bringing forward the next milking session. Natural circumstances allow demand feeding by mutual consent of mother and offspring, which may overcome problems of incomplete gland emptying at the previous milking. In the contrary situation of the machine-milked dairy cow there is no doubt that effective gland emptying is an essential part of a successful parlour routine and maximization of milk yield.

## Nursing and sucking behaviour

In behavioural terms the cow is rather undemonstrative and therefore more difficult and less rewarding to observe. The extrovert pig, on the other hand, shows with clarity many aspects of the physiology and psychology of milking (see Fig. 3.3, p. 36).

Sows suckle their litters approximately hourly, often being roused from slumber by the attentions of one or two members of the litter which have awakened early. If the sow is willing to be sucked she will roll on her side and expose both rows of teats. Phase 1 of the suckling is comprised of the piglets jostling for position on the udder while identifying and retaining their own specific preferred nipple. This phase may last for 20 to 60 seconds. Nosing and butting the mammae with vigorous up and down movements of the head herald Phase 2, which again lasts for about 30–40 seconds. This phase

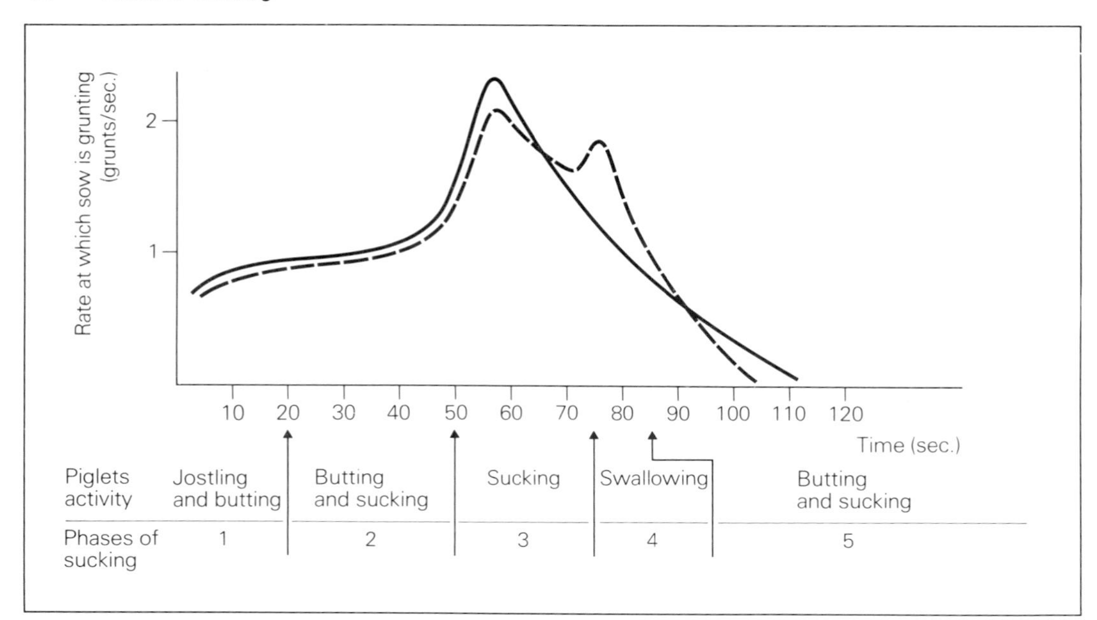


**Figure 3.5** Rhythmic grunting of sows during the course of a bout of suckling. The solid line shows a single peak in grunt rate and the broken line a double peak. The timing of the increase in rate of grunting in relation to milk flow (Fig. 3.3) would suggest that the peak (or peaks) may indicate the release of a burst (or bursts) of oxytocin from the posterior pituitary into the blood stream. Piglets may be seen to swallow milk about 25 seconds after the grunt rate begins to increase

stimulates oxytocin release and the piglets may get a little milk by passive withdrawal from the cistern. The piglets may then be observed to go quiet for about 20 seconds during Phase 3, sucking on the teats with slow mouth movements of about one 'draw' of large amplitude upon the teat every second. Milk flow (Phase 4) lasts for a mere 10 to 20 seconds and may be recognized by the piglet's rapid mouth movements (about 3 per second) of smaller amplitude while they stretch their necks, flatten their ears and swallow 30–50 ml of milk. The final phase (Phase 5) is a brief return to drawing on the teat with slow sucking movements and then vigorously butting the mammae. This phase is usually terminated by the sow rolling onto her udder and covering her nipples.

The sow herself has meanwhile been grunting in a characteristic rhythmic vocalization pattern of about one grunt every second. At around the beginning of Phase 3 of the suckling, this rate of grunting escalates dramatically to about two grunts per second and then subsides down to one grunt per second by Phase 4, and finally the sow will stop grunting sometime during Phase 5 (Fig. 3.5). Not uncommonly two peaks in grunt rate may be heard, the second often occurring around the beginning of Phase 4. It has been observed that not every nursing given by a sow will result in the release of milk, and such unsuccessful sucklings are characterized by the absence of Phase 4 from the routine and a failure on the part of the sow to discernibly increase her rate of grunting. It would appear from the timing and characteristics of the vocalization pattern of the sow that the increase in grunt rate is a ready external indicator of oxytocin release from the posterior pituitary.

# Machine milking

# 4

Efficient machine milking requires fast and complete milk removal without damage to the mammary tissue. Fig. 4.1 shows a cow's teat in a milking machine teat cup, while Figs 4.2 and 4.3 illustrate how the teat cup is set up into the conventional mechanics of a milking parlour. There is a constant vacuum of 330–380 mmHg (45–50 kPa) applied at the inside of the liner to the teat end. The pressure at the outside of the liner alternates from vacuum (380 mmHg, 50 kPa) to air pressure (760 mmHg, 100 kPa). Alteration of vacuum to air (*vac* to *air*) in the outer jacket of the teat cup while constant vacuum is maintained within the liner and around the teat causes the liner first to open and then to collapse. When the liner is open (*vac*), the sphincter at the teat end is forced apart and milk flows. A continuing (*vac*) situation causes the teat to be disfigured, enlarged and forced against the liner wall. Excess blood flow engorges the teat, ruptures the small vessels and causes haemorrhage. Tissue damage and mastitis follow. When the teat sphincter tires it closes off the orifice at the teat end and milk flow ceases. When air is introduced into the outer jacket (*air*) the vacuum at the teat tip causes the liner to collapse and squeeze the teat. The liner collapses in a reverse direction to the sucking action of the young; that is, from the bottom upwards. The collapsing liner forces milk up the teat and back into the cistern, milk flow ceases and the teat end is squeezed, blood vessels become congested and circulation is reduced. During liner collapse (*air*) the teat is said to be given the chance to recover from the milking (*vac*) phase. Whether

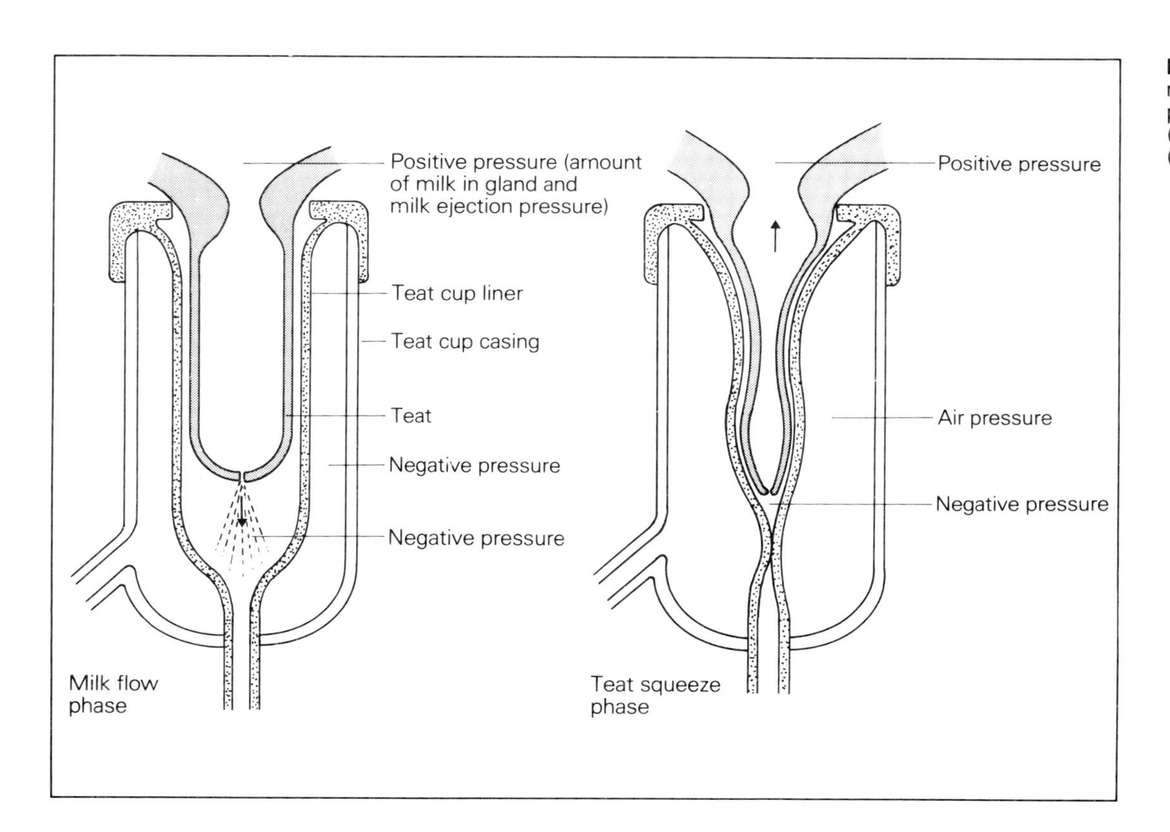


**Figure 4.1** Operation of milking machine teat cup. Negative pressure (vacuum), 380 mm Hg (50 kPa). Air pressure, 760 mm Hg (100 kPa)

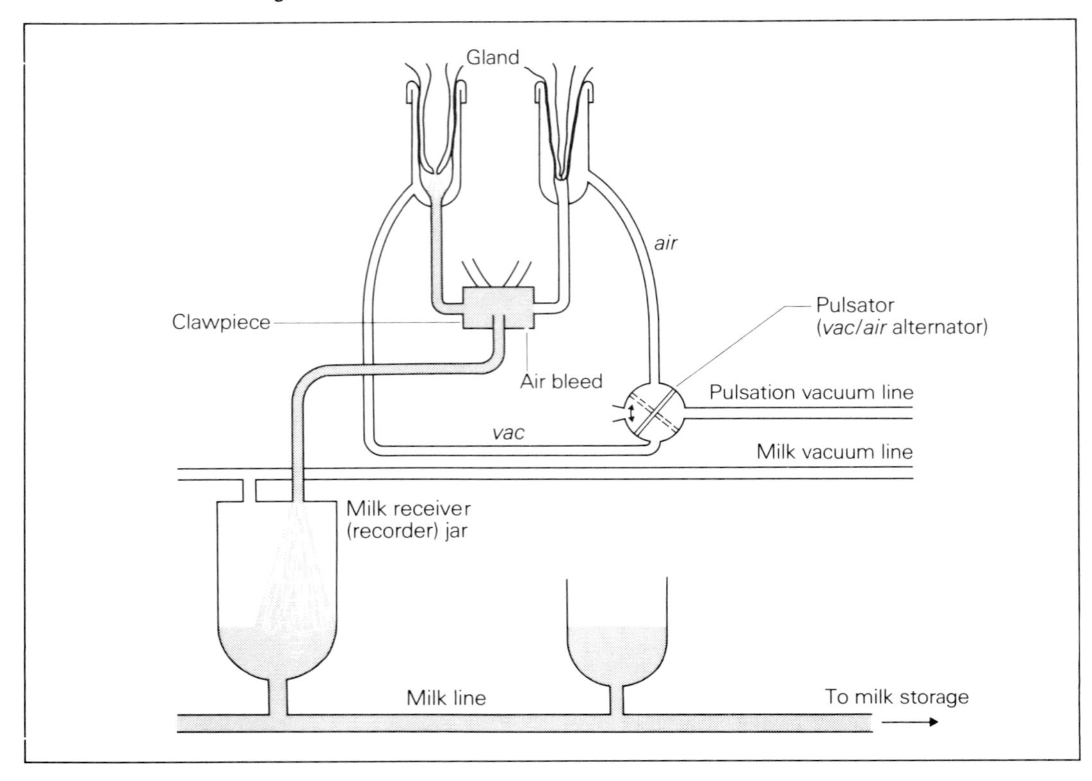


**Figure 4.2** Machine milking set-up

**Figure 4.3** General view of a modern milking parlour in action

the recovery phase is any more restful for the cow than the milking phase must however remain a moot point. The longer the *vac* phase for milk flow, the longer must be the *air* relaxation phase for teat recovery. Increasing the time on *vac*, to increase the milk flow period whilst maintaining the same pulsation ratio of *vac : air*, would require an increase in the total cycle time and a slower pulsation rate. To increase *vac* time whilst maintaining the pulsation rate would require the pulsation ratio to alter and the *air* relaxation time to be reduced. This would increase the proportion of the total cycle time actually spent milking but would reduce the teat recovery time. The longer the *vac* phase and the shorter the *air* phase the greater the chance of fast milking, but also of mammary tissue damage. Usual pulsation ratios (*vac : air*) are 50 : 50 or 75 : 25, but may range from 40 : 60 to 80 : 20.

Whilst milking, milk flows from the udder at the average rate of 1–2 kg per minute, with a maximum rate of about 3–5 kg per minute. Milk does not run freely out of the normal teat, because even after milk ejection there is just insufficient intramammary pressure to open the sphincter at the teat tip. It is the vacuum in the liner which forces open the teat sphincter to bring about milk flow. Milk will flow relative to the size of the orifice made by the opening sphincter and the rate at which the sphincter opens and closes. Milking rate under standard machine conditions is largely governed by sphincter strength and sphincter integrity, and may vary from 0.5 to 5.0 kg per minute. There are large differences between cows in sphincter strength, some cows 'running' easily at low intramammary pressures. Cows with weaker sphincters tend to be those which milk faster, give more milk, and unfortunately are more prone to mastitis. The greater the weight of milk held in the gland cistern the stronger the sphincter needs to be to retain it; contrarily, sphincters of high-yielding cows are under continual pressure and therefore more prone to lapse into weakness. During milking, the positive (push) pressure in the gland cistern and the negative (pull) pressure at the teat end may be added to the intramammary pressure rise following the milk ejection reflex. All these combine to force the sphincter open, notwithstanding the nervous reflex action whose object is to open the sphincter by muscle contraction.

Despite all the forces acting upon the teat end, sphincters of some cows may remain partially closed during the vacuum phase, so reducing the milk flow rate. The sphincters of these cows will only open at full vacuum pressure and will close when any pressure drop occurs. Other cows with weaker sphincters allow the teat orifice to open even at only half full vacuum pressure. Total natural open time for the teat sphincter before the muscles start to close the orifice again is about 0.5 seconds. At slower pulsation rates than 1 pulse per second, the sphincter is already closing before the end of the vacuum phase. At faster pulsation rates the air phase is not of sufficient duration to allow complete sphincter closure and rest before it is snapped open again. Faster pulsation therefore increases milk flow time and puts more strain on the sphincter muscle. Pulsation rates may range from 30 to 70 cycles per minute, but most are of 50–60 cycles per minute.

## Machine pulsation

The possibilities for increasing the rate of milk flow while machine milking will depend much upon the amount of damage that the teat might receive. The wider the pulsation ratio (e.g. 75 : 25 rather than 50 : 50) and the faster the pulse (70 rather than 50 pulses per minute) the quicker and more efficient can be the milking; but the better must be the milking equipment, particularly the liners, and the more exact must be the machine maintenance. Apart from pulsation factors, teat damage also depends upon the design of the cluster and the liners, the material from which liners are made, and the time the machine is in contact with the teats. When milk flow ceases, different teat cup liners leave different amounts of milk in the gland. The bore, length, tension and elastic properties of the liner may make a 10–15 per cent difference in machine milking time. Vacuum levels decreased to above 380 mmHg (50 kPa) and reduced cluster weights (which may range from 2 to 4 kg but are usually about 2.75 kg) also tend to increase the amount of milk left in the gland after machine milk flow has ceased.

There is some tendency for the working vacuum at the teat end to be less than that specified for the equipment. Thus a cluster will, on attachment, exert a vacuum of about 50 kPa, which will reduce by almost 20 per cent as milk begins to flow. As milk flow rate declines, vacuum increases such that at the completion of milking a higher vacuum again acts on the teat. This state of affairs is not ideal, and some pulsator designs may compensate by operating at reduced vacuum (and possibly also at reduced pulsation rate) before the commencement and after cessation of milk flow.

Because the change from *air* to *vac* and from *vac* to *air* is not instantaneous, pulsation response is curvilinear. Various phases of the pulsation cycle are described in Fig. 4.4.1) which shows stylized traces of expected pressure changes in the teat cup. Milk flow will start some time during the *air* to *vac* phase, and milk flow will stop sometime during the *vac* to *air* phase. Cows with strong sphincters will tend to only milk during the full *vac* phase while easy milking cows will tend to milk during most of the *air* to *vac* and *vac* to *air* phases in addition to the full *vac* phase. The effects of pulsation are therefore not simply a function of the time the machine is at full vacuum, but also the proportion of the cycle time spent going from *air* to *vac*, the proportion of the cycle time spent going from *vac* to *air* and the patency of the teat sphincter of the cow concerned.

Figure 4.4.2 shows a decrease in the time spent at the *air* phase when the pulsation ratio changes from 50 : 50 (*a*) to 60 : 40 (*b*). The resulting decrease in collapsed liner time is likely to bring about faster milking, but also a possible

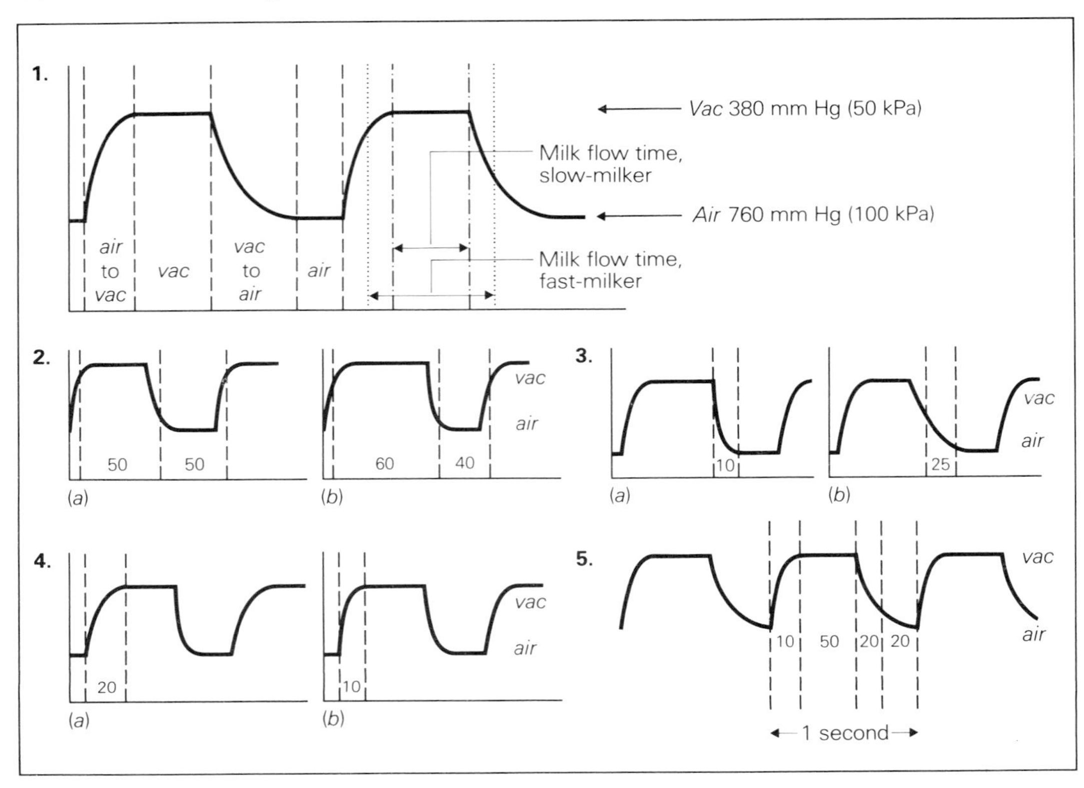


**Figure 4.4** Pulsation. Vacuum pressure changes in the pulsation chamber of the teat cup (see text for explanation)

increase in the amount of milk remaining in the mammae. The possibility of udder damage also increases. The minimum proportion of cycle time on full *air* to allow teat recovery is about 15–20 per cent. Fig. 4.4.3 shows a decrease in the rate of liner collapse; that is, an increase in the *vac* to *air* transition time. The liner is more gentle on the teats in case (*b*); there is less violent milk reversal back from the teat to the gland cistern, and milking time is reduced. Fig. 4.4.4 shows an increase in the rate of the return of the liner from the collapsed position to the open milk flow position; that is, the *air* to *vac* transition time is decreased. In case (*b*) a greater proportion of cycle time is on full vacuum, which means faster and more complete gland emptying. A suggested ideal pulsation pattern is shown in Fig. 4.4.5; this has attributes of a rapid *air* to *vac* sufficient to obtain optimum rate of milking, a slow *vac* to *air* for gentle collapse and an *air* phase of sufficient duration to prevent damage but not so long that time is wasted and the sphincter closes fully. The total cycle time of one second is about optimum in relation to sphincter characteristics and the ratio of milk flow time to rest time is about 65 : 35.

An increasing vacuum inside the liner, at the base of the teat, will effect an increased suction on the teat. Drag at the teat end will increase the downward pressure of milk on the teat sphincter and help to force the sphincter open to allow milk flow. This will increase the rate of milking. Increased vacuum also helps to keep the milking cluster on the cow. However, there is also an increased tendency for the teat cup to crawl up the teat (teat-cup crawl). This will pull into the teat cup some of the mammary tissue, which will close off the teat from the gland cistern and so prevent milk flow, ending the milking prematurely and increasing the amount of residual milk left in the gland. Teat-cup crawl also depends on liner and cluster design, so the better the liner the greater the vacuum that can be tolerated. Vacuum levels of 300–700 mmHg (40–90 kPa) have been tried, while 380 mmHg (one-half atmosphere, 50 kPa) is considered optimum (Fig. 4.5) in achieving a balance between milking rate, tissue damage and completeness of gland emptying. Three hundred and thirty (44 kPa) rather than 380 mmHg (50 kPa) has been found to result in less residual milk, but the rate of milk flow is rather too slow.

The stability of the vacuum is also of importance; fluctuating vacuum will reduce the milking rate. Even at the correct vacuum level, fluttering pressure at the teat cup such as would result from leakage or an inefficient machine, results in a pulsation pattern which, in contrast to that of Fig. 4.4.1, has a long *air* to *vac* phase, a short *vac* to *air* phase and a low proportion of the cycle time actually spent milking.

## Stimulation

An effective milk ejection reflex is needed to achieve fast, efficient milking. Stimulation of the reflex may be achieved by

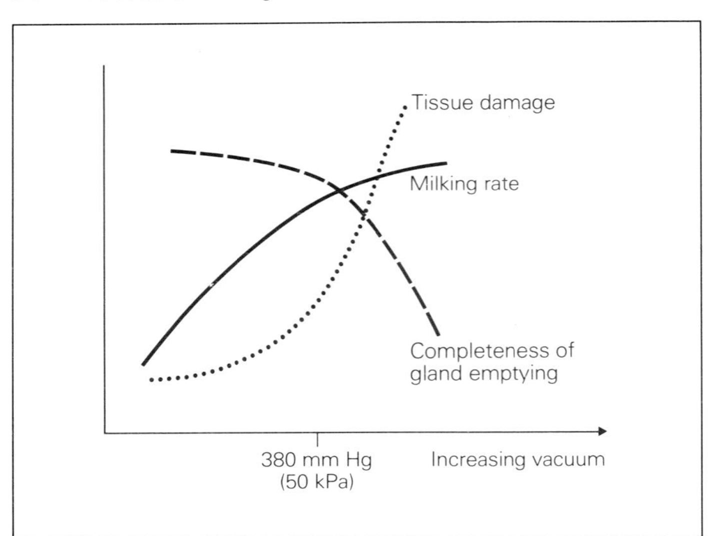


**Figure 4.5** Effects of vacuum level on efficacy of machine milking

conditioned triggers (via eyes, ears and nose) or by tactile activation of nerves in teat and mammary tissue. Tactile stimulation may be achieved by the application of the milking machine itself, or by manual massage prior to application of teat cups, usually when washing. Some experiments have suggested that thorough massage for 30–45 seconds at washing, half to one minute before teat cup application, would give a 10–20 per cent increase in yield and a similar increase in milking rate. Some cows were found to require more stimulation than others, and it appeared that the need for stimulation increased as lactation progressed. It is of course an important part of the massaging technique that the gland is manipulated no earlier than half a minute before teat cup application, or the beneficial effects of good milk ejection reflex would be lost. The effects of manual stimulation will include all the positive benefits of an efficacious oxytocin release, but is expensive in time, unless a part of normal hand washing and not an addition to it. In a modern parlour, hand washing has now been replaced by more efficient methods of cleaning the mammae, such as the hose rinse (Fig. 4.6). Forty seconds of stimulation time would in this case represent an increase of some 15 per cent in the time used for attending to the cow, which would counterbalance any time savings that might have accrued through an increase in rate of milk flow.

There is no doubt, however, that when a cow is familiar with a stimulation at washing and when this is abruptly omitted, the yield will be reduced. It is important to distinguish between the negative effects of loss of stimulation by cows used to a routine which includes it, and the absence of stimulation in the normal routine of a cow's milking life. Account should also be taken of other results from the UK and Eire which have shown no differences in yield, milk flow time or peak flow rate between cows not stimulated and cows

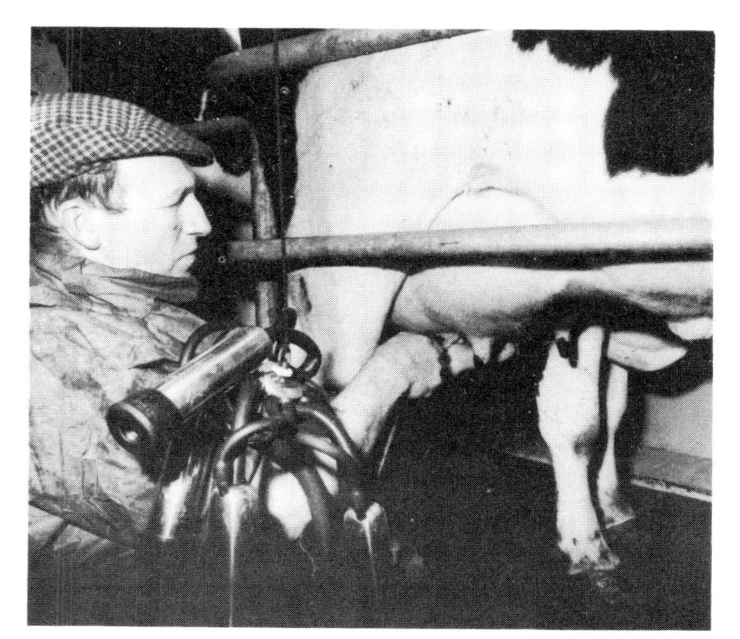

**Figure 4.6** Hose-washing the udder in preparation for machine milking

stimulated by thorough massage at washing 45 seconds or so before teat-cup application.

Given an effectively designed milking machine, a suitable parlour routine and proper handling, application of the teat cups themselves should provide a sufficiently effective tactile stimulation for the milk ejection reflex. In this event there will be a half to one minute lapse between application of the machine and alveolar contraction, but during this time passive withdrawal of cisternal milk will be taking place. It is pertinent that some 40–50 per cent of milk in the mamma of a dairy cow may be available for passive withdrawal. The half minute or so lapse between application of the machine and milk ejection may be particularly beneficial to high-yielding cows, in as much as their cisternal pressure will be reduced, so at milk ejection alveolar milk will flow more readily from the small ducts. Further, an increased proportion of oxytocin activity time is used for withdrawal of alveolar milk, hence there is less likelihood of milk still being present when activity is lost, which will also help complete gland emptying. On the other hand, for cows not at their peak lactation yield, it will be recalled that domestic cows have a large cistern which can accept alveolar milk as it is secreted, so a high level of secretion may be achieved before intramammary pressure slows down milk synthesis. Although faster milking cows have less need of large cisterns, it follows that a large cistern will more readily allow acceptance of milk as it is ejected, so that milk can pass into the cistern for passive milk removal before the effects of oxytocin are lost.

The buffalo (*Bubalus*) has a small cistern and the milk ejection reflex is more difficult to elicit. There can be a latent period between stimulation and milk ejection of two minutes or so. To machine milk buffalo, extra massage is often necessary and the calf may also need to be present. On their home ground, however, buffalo have the potential to yield as much milk and more butterfat than many *Bos taurus* types. There is undoubtedly a general problem with Zebu and buffalo consequent upon the importance of the milk ejection reflex and the apparent unwillingness of these cattle to give their milk away to mankind. There is considerable dispute as to the real nature of these problems. In some cases it appears that at the very least the calf must be present before milk can be withdrawn while in others machine milking to commercial parlour routines can be satisfactorily operated. Different experiences will depend upon the strain or cross of animal used, and upon the conditions imposed upon the animal during the milking. The more important the rôle of oxytocin the more sympathetically must the animal be handled.

## Stripping

Machine stripping is the manual manipulation of the udder whilst the machine is still attached, but after machine milk flow has apparently ceased. About 0.5–1.5 kg of residual milk is invariably left in the gland cistern. This has no effect on the lactation yield and there would be no benefit from attempting to remove it. Higher levels than this of residual milk remaining in the udder may contribute to a faster build up of intramammary pressure and so reduce the secretion rate of high-yielding cows. The objective of stripping is to ease residual milk out from the gland into the teat cup and to possibly bring about a further oxytocin release or mechanical excitation of myoepithelium. The cluster is usually pulled down to prevent teat-cup crawl. Increasing the weight of the cluster assembly to 4–5 kg can reduce the need to machine strip but the cluster is more liable to fall off. In some cases the operative, handling a cluster of 2.5–3 kg, may place a 0.5–1.5 kg stripping weight upon the claw of the teat cluster after 75 per cent or so of the expected yield of that milking has been withdrawn. Where machine stripping is an accepted part of the parlour routine it is usually started when the flow rate of milk from the glands has fallen to about 0.3 kg in 30 seconds; the cluster would then be removed when flow rate has fallen to 0.1 kg in 30 seconds (the average rate of flow during milking is in the region of 1 kg in 30 seconds). A distinction should be drawn between a routine check for gland emptiness (Fig. 4.7), which should take about 5 seconds, and machine stripping, which may take 30 seconds or more. Prevention in this case is therefore better than cure, and to ensure effective milk removal in the first place is much preferable to the inclusion of machine stripping into the routine.

Although it may be necessary to ensure that the gland is

**Figure 4.7** Checking that milk flow has ceased. The cluster is gently pulled down

properly emptied before the machine is removed (this necessitates the operative weighing down on the cluster to counteract teat-cup crawl and giving a quick squeeze to each mammae), additional time spent coaxing out the last drops of milk are likely to be wasted; indeed, the extra machine-on time will only increase the likelihood of mastitis, not to mention the frustrating habit that some cows have of holding back their milk for the specific purpose of releasing it later to the stripping procedure when stripping has become a conventional part of the routine. Stripping will bring about its own indispensability; whereas if never started it may never be needed. When significant quantities of milk remain in the gland after milk flow has stopped this is due to ineffective milk removal and it is probable that either the routine, or the machine, or both is at fault.

If the machine is removed when, in the absence of any interference, milk flow rate has decreased to 0.2 kg per minute, the maximum quantity of milk which could be obtained by active stripping should be only about a further 0.25 kg. Omitting stripping from the routine is unlikely to cause a loss in production of more than 5 per cent of the lactation yield, while it may take up to 20 per cent of routine time. This is worth about 10 cows/man hour of parlour time. The cost of time allowed is greatly in excess of the value of any milk which might have been removed. Machine stripping is also likely to increase the incidence of mastitis because it lengthens machine-on time. As it is, cows are milked on average for about one minute longer than is necessary after significant milk flow has ceased. Automatic cluster removal devices

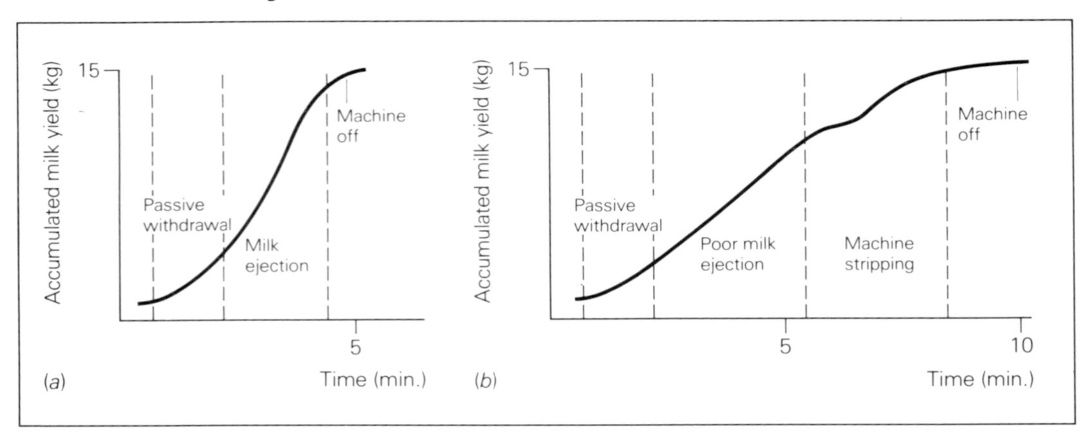


**Figure 4.8** Milk flow rates: (*a*) effective milking, maximum flow rate 5 kg/min., average flow rate 3 kg/min.; (*b*) ineffective milking, maximum flow rate 2 kg/min., average flow rate 1.5 kg/min.

triggered by a minimum milk flow rate from the glands may go some way towards alleviating this problem. Example milk flow rates for effective and ineffective milking routines are shown in Fig. 4.8

In addition to all the pitfalls of machine milking elucidated above, there is the apparently insoluble problem of unequal distribution of milk between the fore and rear quarters of the udder (the ratio is about 40 : 60). Unless the rear glands milk faster, then the use of a single machine on all four mammae must result in either overmilking the fore glands or undermilking the rear.

Finally, there is the problem of the effects of teat structure itself upon the time spent milking. Over and above the marked influence that both the effectiveness of the milk ejection reflex and the time allocated to stripping have upon milking rate, individual mammae may show characteristic milk flow rates. These are almost entirely due to the size of the teat orifice and the strength of the teat sphincter. The flow of milk from teats which have been cannulated, and are thereby of equal bore, is at about the rate of 1.25–1.5 kg/minute for all teats; while uncannulated teats flow at variable rates of 0.5–1.5 kg. The effect of sphincter strength upon flow rate clearly interacts with the mass weight of milk awaiting to exit. Thus, for any given sphincter strength, cows giving higher yields will milk faster. Equally the greater the pressure on the teat end, resulting from the milk ejection reflex, the faster will be the flow of milk.

## Mastitis

Mastitis is encouraged by machine milking, particularly its faults, and may be manifested in clinical or sub-clinical form. In the clinical form, there may be considerable milk losses from infected mammae. About 25 per cent of dairy cows are so infected, as compared to 2 per cent of cows naturally suckled. In the sub-clinical form the loss of milk is difficult to measure from individual cows, but overall loss is considerable. About half the cows in UK have an infection of sub-clinical mastitis. The level of infection will range from insignificant to catastrophic. Average milk losses have been estimated to be around 500 kg per cow per year. Increasing mammary infection causes an increase in the number of cells found in milk; the lymphocyte and leucocyte numbers rise, and so also do the numbers of epithelial cells, whose loss rate is exacerbated. Milk from uninfected dairy cows usually contains about 0.25 million cells per ml of milk, most of these are likely to be cells from the secretory epithelium, discarded in the normal course of events of cell turnover. Rates of loss of cells above 0.25 million per ml is taken to indicate chronic mastitis. The average cell count for all cow milks is 0.4–0.5 million. Calculated losses of milk yield in relation to the cell content of milk is shown in Table 4.1. The estimate that only 25 per cent of all herds in the UK have counts less than 0.3 million, while many could well have cell counts greater than 0.70 million, is indeed sobering; particularly as many of the cells found in milk are spent epithelial cells. The extent to which prematurely lost secretory tissue can be regained will depend upon the stage of lactation and the rapidity and completeness of the cure. The following factors are some of those which have been shown to be predisposing to mastitis: too low a vacuum reserve, too high a vacuum level, fluctuating vacuum, excessive vacuum time, insufficient air time, treat-cup crawl, slow milking, over milking, stripping, excessively strong teat sphincters causing slow milking, excessively weak teat sphincters causing drippy cows and ready infection, poor liner design and inadequate teat hygiene. As a large majority of milking machine plants and parlour routines are faulty to some

**Table 4.1** Losses of milk in relation to number of cells found

| Mastitis rating | Million cells per ml of milk | Kg loss of yield per cow year |
|---|---|---|
| None | 0.25 | — |
| Subclinical present | 0.50 | 300 |
| Subclinical widespread | 0.70 | 700 |
| High level of infection | 1.00 | 900 |

degree in some of these respects, it is hardly surprising that dairy cows are susceptible to mastitis. There is certainly no reason to believe that milking machine design is unimpeachable. Suckled cows given excess calves give more milk than if machine milked. The calf is the most perfectly designed milking machine, and the machinery we have to hand at present has far to go before it can mimic the calf, either in action or end result.

## Frequency of milking

The frequency with which cows should be milked to obtain maximum yield will depend upon the amount of milk in the gland in relation to the oxytocic activity time, and the time taken since the last milking for intramammary pressure to reduce secretion rate. These factors relate to the milk storage capacity of the gland, the potential rate of milk secretion, and the effectiveness of the milk ejection reflex in clearing the gland of milk. An increase in the frequency of milking may therefore have beneficial effects when yields are very high, when cistern size is small in relation to yield, when milking rate is slow, or when myoepithelium contraction time is limited.

Calves suckle every 3–5 hours, dairy cows are milked twice, or at most three times, daily. The crucial question in relation to the dairy cow is therefore whether or not the cistern and duct storage spaces are large enough to hold all the milk which could potentially be secreted between milkings. Milk secretion rate is linear for cows of *average* yield for up to 15 hours after milking; after this time, intramammary pressure effects curtail secretion of milk from the alveolar epithelium. Thus, a 12/12 hour interval between milkings may be no better than a 10/14 hour interval. This, however, is not the case for cows of high yield, whose milk secretion rates are likely to become inhibited after an interval between milkings of some 9 hours or so.

The short-term response to milking an average dairy herd three times daily at peak yield is usually some 5–15 per cent increase in yield. Longer term responses have been estimated at 5–20 per cent increase in yield; the long-term response may be due to additional stimulation of hormones influencing lactation maintenance. Increasing milking frequency is likely to be most effective for cows with limited storage capacity but with active epithelia prevented from further milk production by back pressure. The higher the potential daily milk yield, the more likely is three times daily milking to be worthwhile.

## Parlour routine

It is possible to achieve a milking rate of 60 cows per man hour if the cows are relatively clean, the cows are average to fast milkers, and neither manual massage nor machine stripping are practised. It is helpful to the routine if either all, or at least

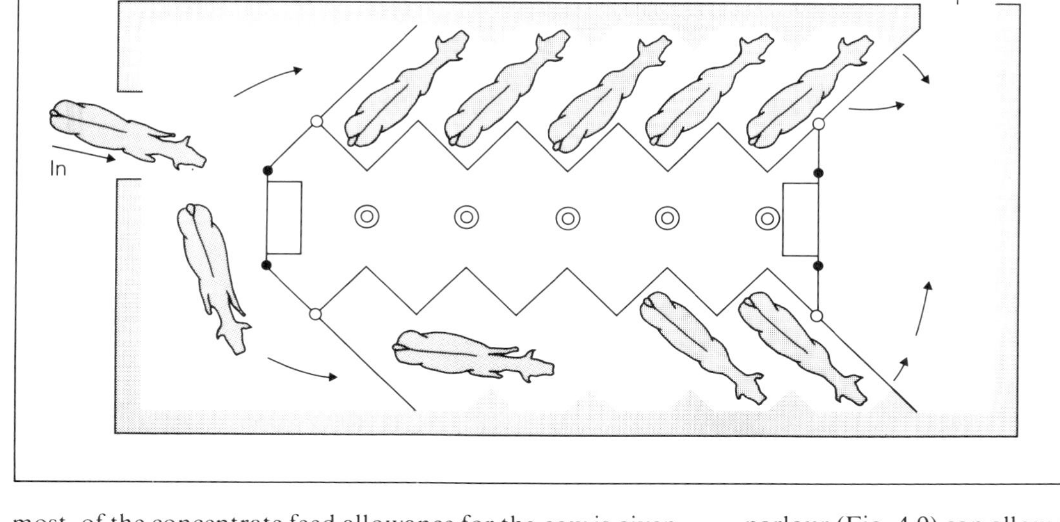


**Figure 4.9** Five-unit/10 stall herringbone parlour. One man can milk 60 cows per hour. The cows are dealt with in batches of 5. While the left-hand side is being milked the right-hand side is let out and 5 fresh cows let in, the foremilk is examined and the glands washed and dried. The operative returns to the left-hand side (which is still milking), checks the glands to be empty and transfers the units one-by-one across to the right-hand side. The teats of the left-hand cows are dipped and they are let out. A fresh batch of 5 cows is let in to the left-hand side

most, of the concentrate feed allowance for the cow is given outside the parlour. Further benefits would also accrue from discarding from the herd cows which are liable to mastitis or which milk slowly. Fast milking cows take 3–4 minutes to milk, average cows 5–6 minutes and slow milking cows 7–8 minutes. An operative with 5 units in a 10-stall herringbone parlour (Fig. 4.9) can allow 5 minutes of unit-on time per cow. This is usually enough time for all but the highest yielders. The operation time should be about 1 minute per cow (as depicted in Table 4.2 and Fig. 4.10) to be compatible with one man handling 5 units. Unless the herd is slow milking (and if so, why?) more than 5 units/man will simply result in

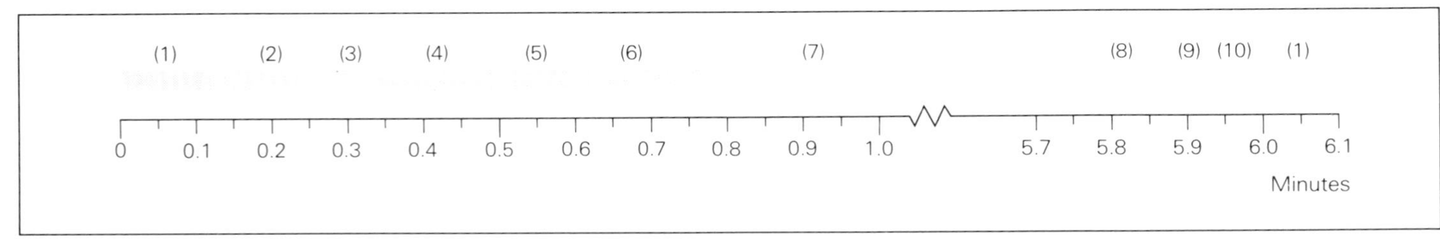


**Figure 4.10** Analysis of time taken for the various operations of an efficient parlour routine (minutes per cow).
(1) Let cow out.
(2) Let cow in (and feed if concentrates offered in parlour).
(3) Take foremilk.
(4) Wash udder (hose rinse with soft brush).
(5) Dry udder (paper towel).
(6) Attach teat cups of machine.
(7) Milking (5 minutes).
(8) Remove machine from udder (check completely milked).
(9) Dip teats in disinfectant.
(10) Miscellaneous other tasks.
Total time: One minute's operation time; 5 minutes' milking time; 6 minutes' total.

overmilking. In most modern parlours there is one unit to *each* stall, in this event the units remain idle for about half the time (see Fig. 4.3, page 45). The principal advantage of one unit for each stall (rather than one between two) is that the machine-on time can be more flexible, thus allowing for slow milking cows and high yielders. Also important is the spin-off benefit of an individual receiving jar for each cow; this can be located *below* each stall, and this position makes for much more even vacuum characteristics with the advantage of faster milking (Fig. 4.11). In the 1 unit/2 stall set-up the receiving jar must be above cow height and in the middle of the parlour to allow the cluster to be swung across from one side to the other.

A device for automatic cluster removal, such that the machine is withdrawn without assistance when milk flow has

**Table 4.2** Operation times (minutes) in a 5-unit, 10-stall parlour

| | |
|---|---|
| Let 5 cows out, let 5 cows in, feed automatically | 1.00 |
| Foremilk, spray wash, individual paper towel dry, 5 cows | 1.75 |
| Change over units, off 5 cows and on to 5 cows | 1.75 |
| Dip teats of 5 cows* | 0.50 |
| Total time for 5 cows (minutes) | 5.00 |

*The teat dip can be a dilute solution of disinfectant such as hypochlorite or iodophor together with an emollient such as glycerol or lanolin to keep the teats soft.

**Figure 4.11** A 16-unit/16-stall herringbone parlour to cope with a herd of 200 cows

ceased, can save about 0.15 minutes of routine time. A further 0.05 minutes can be gained by use of automatic teat spraying with disinfectant at the end of milking instead of a teat dip. If the cows enter their stalls without help, and if the animals are not fed in the parlour then another 0.25–0.30 minutes can be saved. The essential operations are now reduced to items 3–6 enumerated in Fig. 4.10. These tasks take about 0.5–0.60 minutes or so, and allow a potential throughput of 100–120 cows/man hour.

The rotary parlour (Fig. 4.12) may, if properly managed, increase the rate at which cows can be put through the milking routine. The parlour is moved round by a motor activated either by the operative or automatically (or both). As each cow reaches the exit she walks off the stand and vacates a space for the waiting cow to enter. If the milking time for each cow is about 5 minutes, then the number of milking stalls (with one unit/stall) required on the parlour will be about 10, plus the two stalls used up at the entry and exit points. Fig. 4.12 shows a 12 unit/12 stall rotary parlour; 14/14 set-ups are also common.

Parlours for two full-time men, or to allow for different milking times, and/or different essential operation times, would, of course, require a different number of milking places.

All parlour routines should incorporate features to ensure satisfactory milk ejection, fast milking and freedom from mastitis. The cows must feel comfortable and secure and be willing to release their milk. The routine must be smooth,

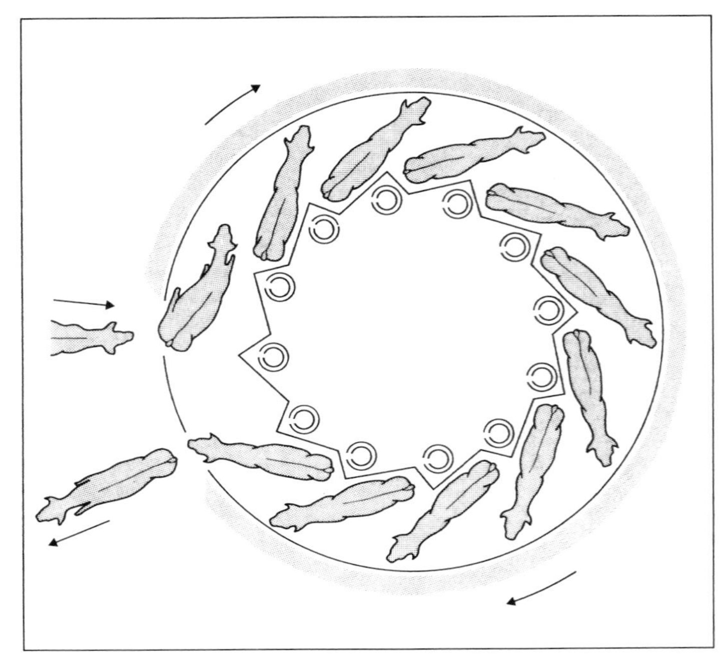

**Figure 4.12** Twelve-stall/12-unit rotary herringbone parlour

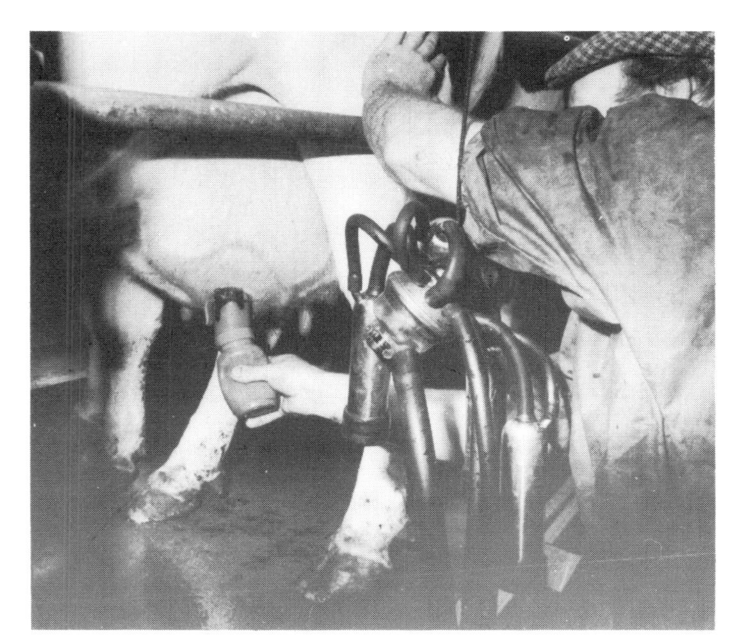

**Figure 4.13** After the cluster has been removed, the teat is dipped in an antiseptic solution

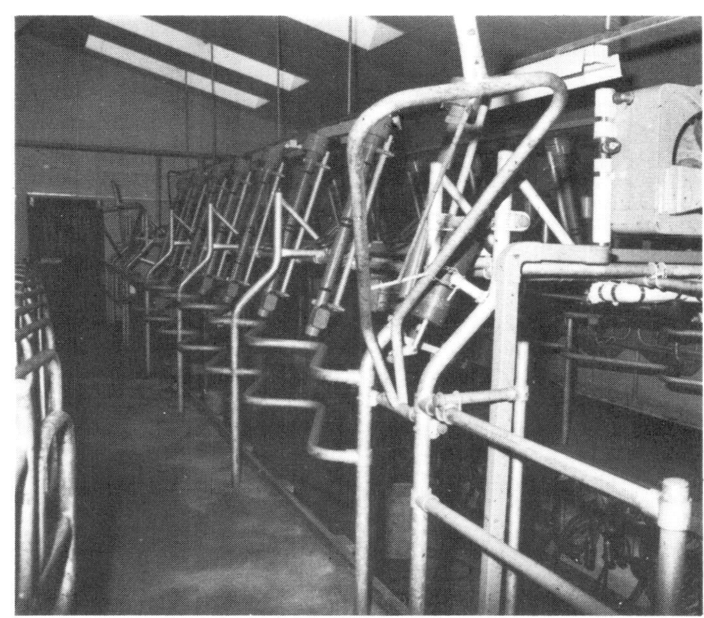

**Figure 4.14** Automatic cluster removal kit

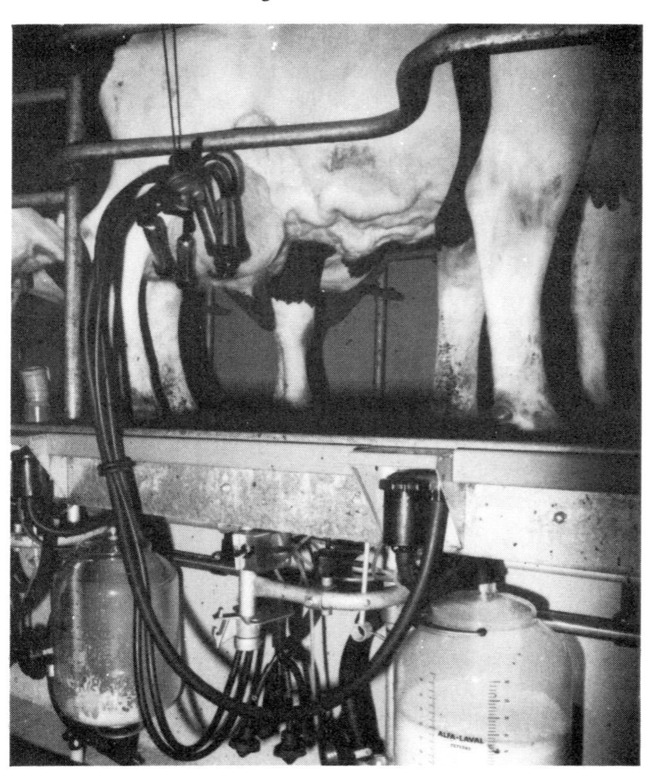


quiet and regular, the machines effective but gentle and the operatives sympathetic. Hoses and rinsers are preferable to cloths for washing, and disposable towels best for drying; teat hygiene can be assured by dipping the teats in disinfectant following milking (Fig. 4.13).

Automatic devices for removing clusters (Fig. 4.14) must be carefully adjusted. As milk flow from many cows is variable, trigger-happy cluster removers may take the machine off before the cow has finished milking (Fig. 4.15). Conversely, a failure to remove the cluster at the right time results in overmilking. It is therefore necessary that features such as a pre-set minimum machine-on time, say to 2 or 3 minutes, are incorporated. Final removal should *not* be determined on the basis merely of a pause in milk flow, but rather when the flow rate has tailed to reach a given rate, of say 0.2 kg/minute, for a predetermined time period, of say 0.25–0.5 minutes.

A cluster remaining on the udder after milk flow has ceased can only do harm; both to the mammary tissue and the efficiency of the parlour routine. Automatic cluster removal is therefore a considerable asset to effective milk removal.

The milking machinery itself must be kept clean by sterilization after each milking. Above all the milking equipment must be operated at top performance level – this means regular inspection and maintenance.

**Figure 4.15** Automatic cluster removal in action

# Characteristics of lactation yield

# 5

An obvious point, although one often overlooked, is that lactation yield can only be as great as the amount of secretory tissue available to synthesize milk, and the ability of the animal to ingest the food from which come the raw materials of milk manufacture. It is not surprising that a cow will give more milk than a goat, nor that weight for weight, the human female can be as good a milker as the dairy cow. Neither is it unreasonable to find that when breeds of dairy cow are compared, the heavier breeds tend to give more milk (Table 5.1). Nevertheless, there are wide differences between strains within breeds which are not accountable for in terms of body weight; the best strains of Ayrshire cows will often outyield average Friesians. Genetic selection within breeds for high-yielding strains will, almost undoubtedly, bring about some yield improvement. The permanent benefits which can accrue, albeit gradually, from breeding better milking cows should not, however, obscure the very large and dramatic affects on yield which can be effected by changes in cow management and nutrition. Neither should it be overlooked, that from an

**Table 5.1** Influence of size upon milk yield

| Breed | Approximate cow weight (kg) | Approximate lactation yield (kg) |
|---|---|---|
| Friesian | 600 | 4500 |
| Ayrshire | 500 | 4000 |
| Dairy Shorthorn | 500 | 4000 |
| Channel Island | 400 | 3500 |

individual producer's point of view, the quickest way to improve the yield of one breed of cows might be to change to another.

Both daily yield and lactation yield are affected by the vigour and effectiveness with which the milk is removed. It is almost axiomatic by the nature of things that yield is a challenge : response, demand : supply phenomenon. If more milk is asked for, more milk will be provided. Table 5.2 demonstrates the principle by showing the average milk yield of a sow in relation to the number of piglets in the litter; the larger the litter, the greater the yield. The weight of the young is positively related both to their ability to draw milk from the udder and to their requirements; thus the yield rises to a peak at 3 to 4 weeks as the litter grows (Fig. 5.1 and Table 5.3). Decreasing yield from 4 to 8 weeks is associated with an increasing intake of solid food by the young. With naturally suckled piglets, nutritive independence without growth retardation is possible from 6 weeks of age and probable by 8 to 9 weeks. With the cow, strength and capacity of the young to withdraw milk dramatically affects the daily yield of milk. The increasing yield as the lactation progresses is a direct function of the ability of the calf to draw ever greater amounts of milk from the gland until the maximum potential yield is reached. The decline in lactation begins as the phase of increasing cell number and cell activity is concluded and the phase of decreasing cell number and decreasing activity begins, so the secretion rate falls away. The decline in milk yield is commensurate with the ability of the calf to feed itself and be independent of mother for its sustenance.

**Table 5.2** Milk yield of sows as influenced by number of pigs in the litter

| Number of piglets in the litter | Milk yield of sow (kg/day) | Milk intake of piglets (kg/piglet/day) |
|---|---|---|
| 6 | 5–6 | 1.0 |
| 8 | 6–7 | 0.9 |
| 10 | 7–8 | 0.8 |
| 12 | 8–9 | 0.7 |

**Table 5.3** Increasing yield of sow as associated with increase in live weight and vigour of the litter

| Age (weeks) | Weight of litter (kg) | Milk yield of sow (kg) |
|---|---|---|
| 0 | 13 | 1 |
| 1 | 20 | 5 |
| 2 | 35 | 7 |
| 3 | 50 | 8 |
| 4 | 70 | 8 |

The milking machine is analogous to an infinitely voracious calf. Fig. 5.1 shows lactation curves for a beef cow, suckled or machine milked. The curve for the suckler cow is similar, not surprisingly, to that of a suckled sow. The use of the milking machine may be contrasted to that of the calf in that the machine reaches its maximum appetite instantaneously and maintains it constantly throughout the lactation. The machine

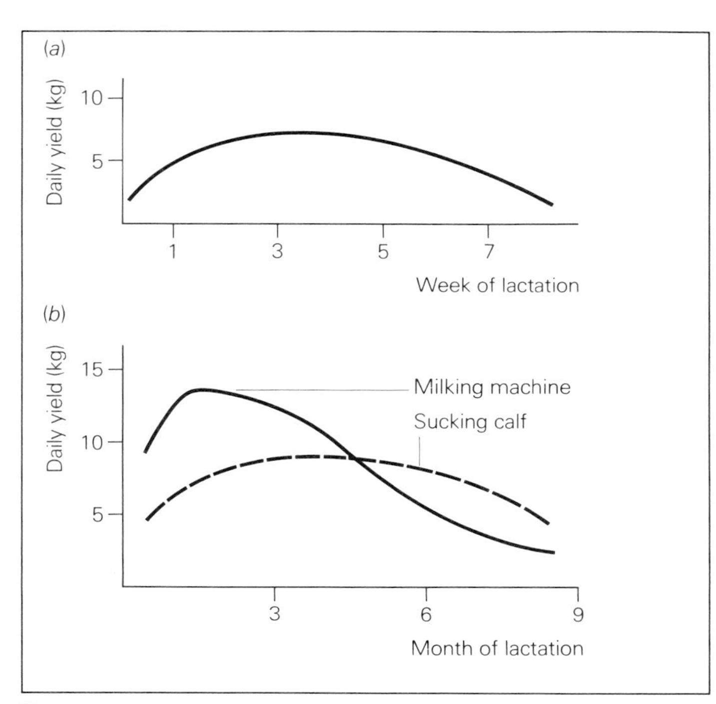


**Figure 5.1** Lactation curves. (*a*) for a sow; (*b*) for a cow

curve reaches its peak earlier than the suckler curve, and the peak is higher. Conversely, the rate of yield decline from peak is more rapid for machine-milking, and the yield tails away earlier. It could be speculated that the use of the machine has not increased the total yield as much as might have been expected, due to reduced persistency compared to suckler lactation. If the peak for the machine curve were to be followed by the persistency of the suckler curve, one would achieve a lactation yield much nearer the potential than is to be found in the case of the more usual dairy lactation depicted by the solid line in Fig. 5.1(*b*).

Figure 5.2(*a*) shows the self-evident point that longer lactations will give greater lactation yields in the form of a classical diminishing curve. A similar diminishing response expresses the relationship between age at first calving and the yield for the first lactation (Fig. 5.2*b*). It is probable that the age at first calving effect is mostly a result of animal weight and condition; these two improving as the animal gets older. Much of the age effect apparent from the figure is likely to disappear if nutrition is improved commensurate with earlier calving. The relationship shown, therefore, should not be taken to be an indicator of the benefits of later calving. On the other hand, the loss of one year's yield of milk together with the extra keep for the cow more than offsets any extra yield there might be in the first lactation following a delayed calving. The most important message is that although it is unquestionably beneficial to calve young dairy cows for the

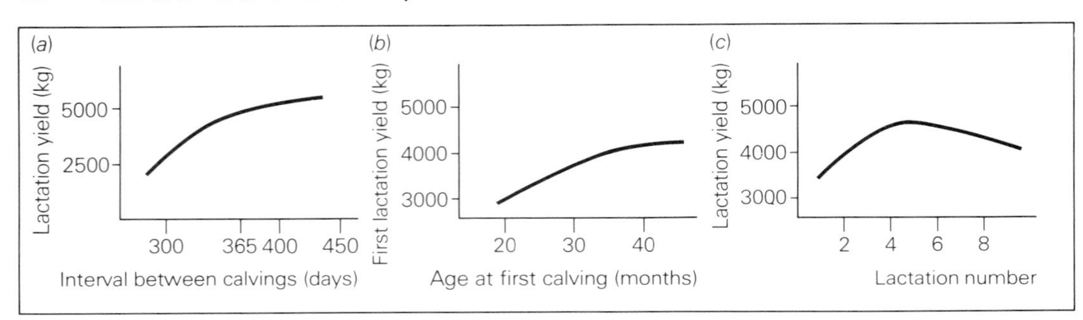


**Figure 5.2** Some factors influencing lactation yield

first time at about 2 years of age under UK conditions, unless the nutrition for the animals is adequate and unless they reach the point of calving in good condition, then first lactation yield will be diminished. Current recommended practice therefore is to mate young females at 15 months of age to calve at 24 months. Respective liveweights for Friesians should be around 340 kg at first mating, 500 kg just before calving, 450 kg after calving and 600 kg at maturity.

As the animal grows, lactation yield increases; then as aging occurs lactation yield is reduced forming a hump-back response for lactation yield to lactation number (Fig. 5.2*c*).

## Seasonality

The 'spring hump' is a colloquial term to describe the phenomenon in UK of a rise in the milk yield when the cattle are turned from their winter quarters and conserved feeds out on to the grass pastures of young spring foliage. The influence of the spring hump on the lactation curve raises yield by 10 to 15 per cent, regardless of the point in the lactation curve at which it occurs. If the hump occurs in mid-lactation, lactation yield is raised by up to 500 kg. The effect probably results from improvement in nutrition consequent upon feeding spring grass in place of winter forage. Fig. 5.3 shows the influence of spring hump upon the shape of the lactation curve of cows calving in January, May and October respectively. The effects of the spring are most beneficial for the winter calver as it is the persistency of the yield which is maintained. For the spring calver there is an increase in the height of the peak, with subsequent rapid lactation decline. The autumn calver catches the spring hump effect too late on in the lactation to do any good; 15 per cent of a low yield cannot be as useful as 15 per

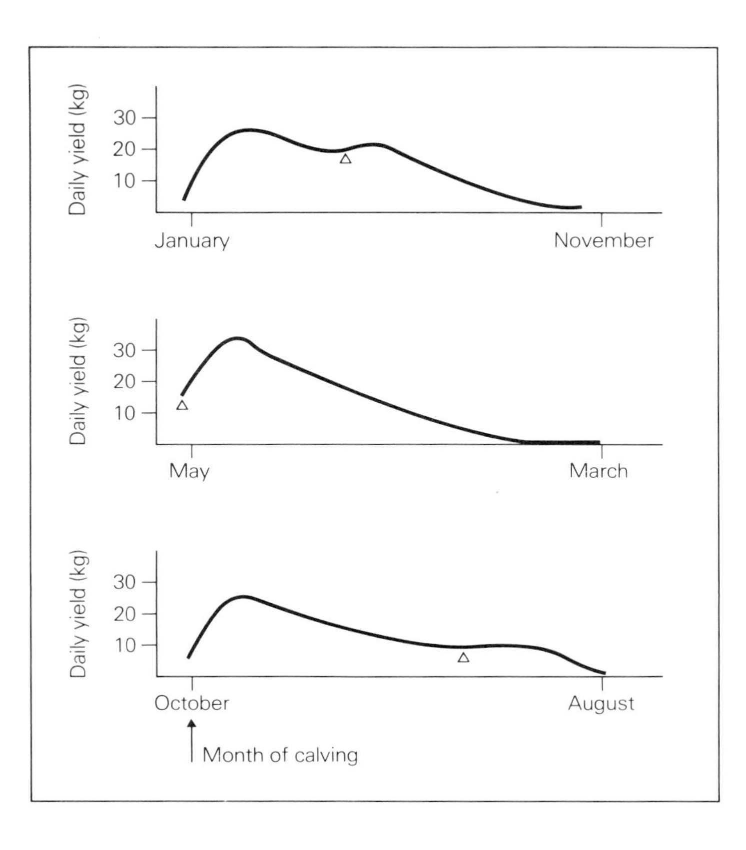


cent of a higher one, and coming towards the end of the lactation the spring hump effects do not last for very long.

In the UK the time of year in which the cow calves also influences her yield. Cows calving between November and February give some 5 per cent more milk than the average, while cows calving between March and July gives some 5 per cent less than the average. This affect appears to be independent of spring hump and of nutritional influences. All things considered, cows calving in the winter between October and January have lactations about 600 kg higher than cows calving at other times of year. This is due to the combined positive effects of both month of calving and spring hump. Seasonality effects are expressed in percentage terms in Fig. 5.4.

## The lactation curve

The changes in daily yield as the lactation progresses are a direct result of changing cell number and cell activity. Despite the many possibilities for the lactation curve of individual cows to be rather variable, the average lactation curves for machine-milked dairy cows do tend to show a number of common characteristics. The curve is a function of time after calving, the general level of production, the height of the peak,

**Figure 5.3** Influence of season of year upon the shape of the lactation curve (spring hump effect). The onset of spring is marked by the symbol △

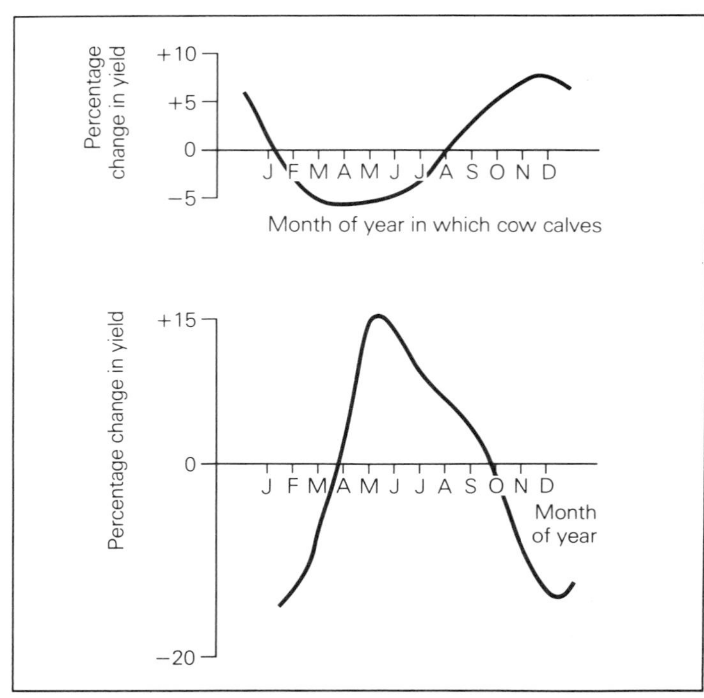


**Figure 5.4** Influence of month of calving and season of year upon milk yield (interpolated from Wood, P.D.P., 1979, *Animal Production*, 11, 307)

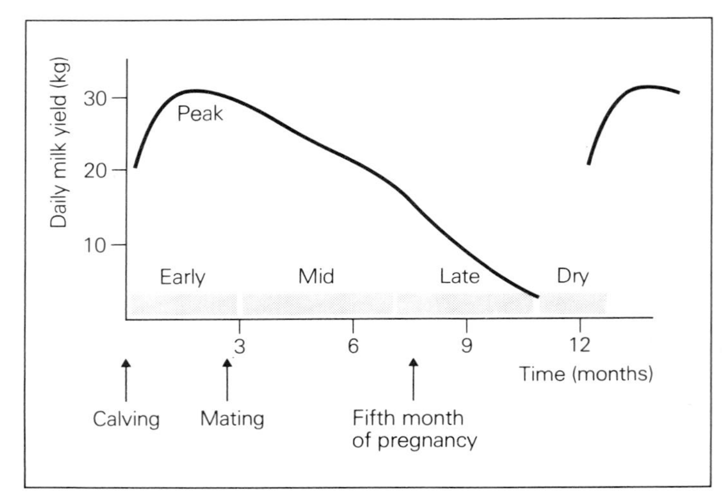


**Figure 5.5** Lactation curve of a dairy cow

the rate of rise to that peak and the rate of decline from the peak. A generalized lactation curve for a high-yielding cow is shown in Fig. 5.5. The drawing shows a curve with a peak yield of about 30 kg. The lactation has a duration of about 320 days, which allows a dry period for mammary regeneration of 45 days and is consistent with one lactation yearly. The curve has been divided into four periods: early lactation from calving to the beginning of the post-peak decline (usually about 12 weeks), mid lactation which covers the phase of gradual yield decline (12–30 weeks), late lactation which usually begins about the time of the steeper decline of the curve (which coincides with the fifth to sixth month of pregnancy) and continues to the termination (30–44 weeks), and the dry period which usually lasts from 30 to 60 days. Peak yield usually occurs some time between 4 and 10 weeks after calving. As a rule of thumb the total lactation yield can be estimated by multiplying the daily peak yield by about 200. For lactations with low persistency a number of 180 may be more appropriate, while for persistent lactations a factor of 220 could be used. Two other useful approximations are that about two-thirds of the total yield occurs in the first half of the lactation and the rate of decline of the yield from peak is at about the rate of 2–2½ per cent per week.

The average lactation curves for machine-milked dairy cows can be quite well described by use of a simple mathematical equation. The possibility that lactation yield could be roughly predicted from relatively little information clearly has many attractive attributes when it comes to planning the management of the feeding and organization of the dairy herd. One equation (suggested by P. D. P. Wood of the Milk Marketing Board) is

$$Y = An^{b}e^{-cn}$$

which, although it looks formidable, is simple to handle with a computer or modern pocket calculator. $Y$ is the average daily yield in the nth week of the lactation. $A$ is a scaling factor fixing the height of the curve and relates to the average production level of the herd and the specific potential of individual cows. $n$ is the week number for which a yield estimation is required. The constants $b$ and $c$ describe the curves respectively up to and after the peak, and e is the base of natural logarithms (2.718). The numerals inserted into the equation must vary with individual circumstances, but one example is given in Table 5.4 for cows in their first and fourth lactations.

**Table 5.4** Some example factors for use in equation to approximate the lactation curve

| Lactation number | $A$ (kg) | $b$ | $c$ |
|---|---|---|---|
| 1 | 16 | 0.15 | 0.03 |
| 4 | 22 | 0.24 | 0.05 |

The table shows how the fourth lactation yield is greater, rises to a higher peak but is less persistent than the first lactation yield. Such prediction equations are now in limited use both at the research and farm management levels. Caution should, however, be exercised in the interpretation of this sort of model. The point is not that a herd should follow the predetermined curve, but once such a curve is set up it could act as a datum; deviations from it observed, and the reasons elucidated. Action, if necessary, can then be taken on an objective basis. A further cautionary point is that

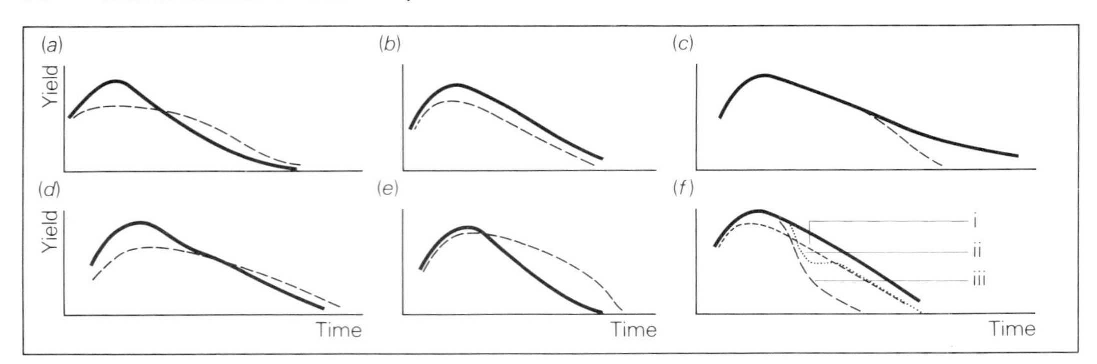


**Figure 5.6** Some factors influencing the shape of the lactation curve. See text for explanation

*mathematically, A, b* and *c* ought to be independent, which *biologically* they are not.

Some of the factors which can cause the shape of the lactation curve to deviate from the expected average are shown in Fig. 5.6. The higher the peak of the lactation curve the faster tends to be the rate of fall (Fig. 5.6*a*). This is likely to be consequent upon the greater loading of the higher peak upon the secretory tissue, causing it to have a more stressful and foreshortened life. Further, the need for the lactation to end within 320 days or so to allow regeneration will automatically force upon a higher peak a steeper rate of decline. Despite this, lactation curves with high peaks are of high yield because the decline is from a higher point (Fig. 5.6*b*); cows may still be yielding measurable quantities of milk when they require to be dried off. The height of the peak of the lactation curve is the greatest single determinant of the total lactation yield. It is because of this that such a large proportion of the effort of dairy farming is put into maximizing the height of the lactation peak. Present restrictions to achieving maximum peak are largely nutritional, and only by the most skilled feeding management can the optimum peak yield be achieved.

Figure 5.6(*c*) shows the effect of the fifth month of pregnancy causing an increase in the rate of yield decline. Figure 5.6(*d*) gives a typical comparison of a cow and a heifer lactation, the broken line lactation having a lower peak and greater persistency. The tendency for the peak to become higher and rate of decline to become faster continues up to the fourth lactation or so. Figure 5.6(*e*) depicts two lactation curves both with the same peak, but one more persistent than the other. The positive advantage of the peak yield being held over mid-lactation is clear, and high lactation yields can only be attained when the lactation curve has both a high peak and shows persistency. There is some argument to the effect that pushing cows to too high and too early a peak will be self-defeating in that the yield tends to fall away disappointingly in mid-lactation. This alternative school believes that the nutritional difficulties of achieving a high peak, together with the possible failure of the yield to be maintained post peak, might make it more worthwhile to be satisfied with a lower peak, but to maintain the yield for longer through mid-lactation. Although this matter is not resolved, it is clear that neither a high peak nor good persistency will alone give high yields; a high peak is needed *together with* good persistency.

## Nutrition

A major factor influencing the milk yielded must, of course, be the supply of nutrients; the milk secretory cells can only synthesize milk constituents if they are given an ample supply of raw materials from the blood. The cow has two sources of nutrients: her own body, which can be broken down to provide for the requirements of milk synthesis (Fig. 5.7), and the food she eats. Given that the cow's body stores are limited in

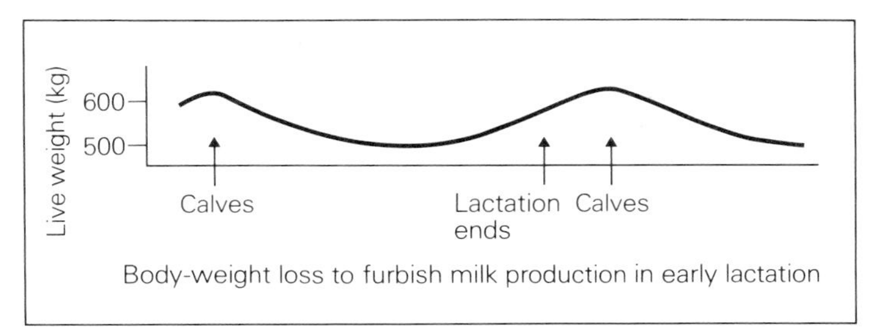


**Figure 5.7** Body-weight changes in the dairy cow

supply, yield of milk tends to be proportional to the amount of food the animal eats. One of the main reasons for bigger cows giving more milk is because they eat more food; and it would be true to say that the most rapid improvement in any dairy cow herd's milk yield could be achieved simply by getting more food down the animals' throats. Within any herd the highest yielding cows need, and tend to have, the biggest appetites.

In the long term, the availability of nutrients (level of feeding) and the condition of the body (level of fat stores) will influence the rate of milk secretion by affecting the height of the peak, the time after calving that the peak occurs, and the post-peak persistency. In the short term, at any time in lactation, a temporary shortfall in nutrients which cannot be buffered by body stores will reduce the rate of secretion.

Decline in yield is a function of previous *actual* yield, not previous potential yield; thus a temporary loss in yield adversely affects the whole of the subsequent lactation. Fig. 5.6(*f*) shows the affect of a temporary reduction in feed supply at peak (i), after peak and showing some recovery following reinstatement of an adequate feed supply (ii), and after peak with no recovery despite refeeding (iii). In no event is the potential yield fully regained. The extent of milk yield loss consequent upon a shortfall in nutrient supply will depend upon the condition of the cow and the amount of energy available from her own body stores. The importance of nutrition as a variable under the control of the dairy farmer, as well as being the most important determinant of yield, merits the much fuller exposition of the subject.

## Lactation length

On the basis of a between-species comparison, natural lactation length in mammals is broadly related to body weight. With mammals kept for purposes of milk production (dairy breeds of cows, goats or sheep), or for purposes of their offspring which are temporarily dependent upon milk (breeding females of pigs, sheep and cattle destined for meat production), lactation length is usually imposed by man. Unlike pregnancy, lactation length can be manipulated.

In the natural course of events, milk synthesizing cells are lost throughout the lactation. But in early lactation the rate of replacement is higher, and in late lactation lower, than the rate of loss. As cells lost are not replaced, the yield reduces until it finally ceases. This is a gradual process and may take 300 to 600 days or more in the non-pregnant dairy cow. Pregnancy itself hastens the end of the current lactation due to the need to refurbish the gland ready for the next lactation. The high-yielding dairy cow is sometimes imperfectly organized and may on occasions still be lactating less than 30 days prior to the forthcoming calving. In this event, nature needs a helping hand and she must be forcibly dried off by stopping milking if her next lactation is not to suffer. The build-up of milk constituents in the mammary gland following cessation of milk removal acts as a negative feed-back to milk synthesizing cells. After synthesis has stopped, reabsorption will take place. The influence of the hormones associated with synthesis, stimulated by the sucking young and the removal of milk, fades in the absence of regular milking.

Unless the cow has prematurely terminated her own lactation following a rapid decline from peak, lactation length of dairy cows is imposed primarily by the impending calving. In domestic species this breeding cycle is, at least in part, under the control of man; which of the three-weekly oestrous periods is chosen for the female to be mated is an active decision of the dairy farmer, not the cow. The influences of pregnancy upon lactation length can be seen in the dairy cow from mid-lactation after which the effect increases

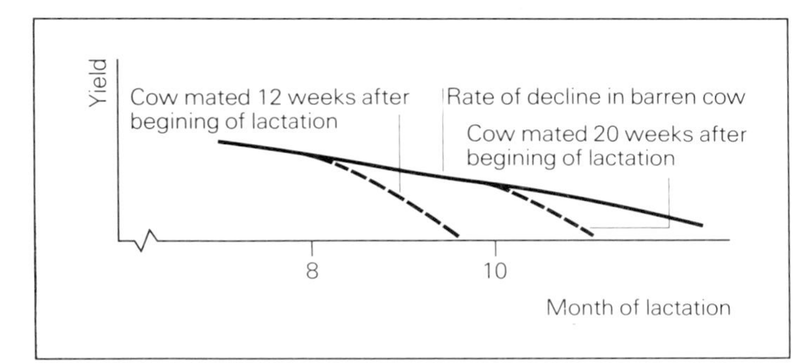


**Figure 5.8** Influence of pregnancy upon lactation length

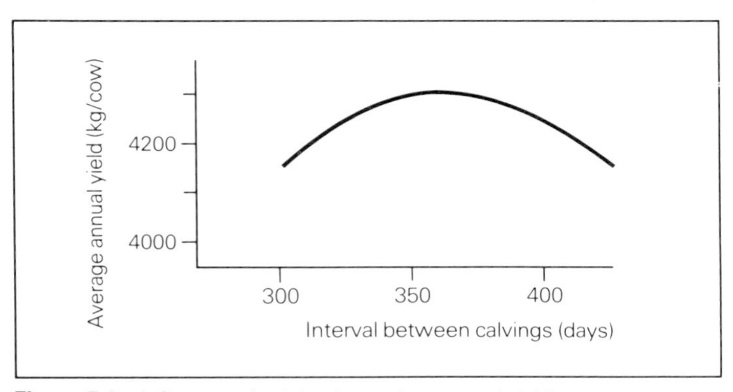


**Figure 5.9** Influence of calving interval on annual yield

progressively (Fig. 5.8). The impetus to have dairy cows mated within 12 weeks of calving follows from the assumption that one calf per year, and in consequence one lactation per year, gives more milk per day on average than do long drawn-out lactations with many days of low yield at the tail end. Optimum lactation length also depends upon the value of another calf against more milk in late lactation. Maximum annual yield appears to accrue if the cows are mated to calve once every 350 to 370 days (Fig. 5.9). Although long lactations are better for individual cow yield, short lactations are better for annual herd productivity.

Most herds do not achieve more than one calf on average in 370 days, and for many one calf every 390 days is probably nearer the mark. Cows tend to slip about one month per year; first calvers in November tend to have their fourth calves in February. In consequence, seasonality of production is rarely maintained. The ideal of a 305-day lactation, a 56-day dry period and 82-day interval between calving and mating (Fig. 5.10) is seldom achieved. The reason for this is that even at best, conception rates for dairy cows are seldom better than 75 per cent and are nearer 55–70 per cent over the 50–80 day period after calving during which the cows must conceive to hold to a 365-day calving interval (Fig. 5.11). The chronic infertility from which most high-yielding dairy cows

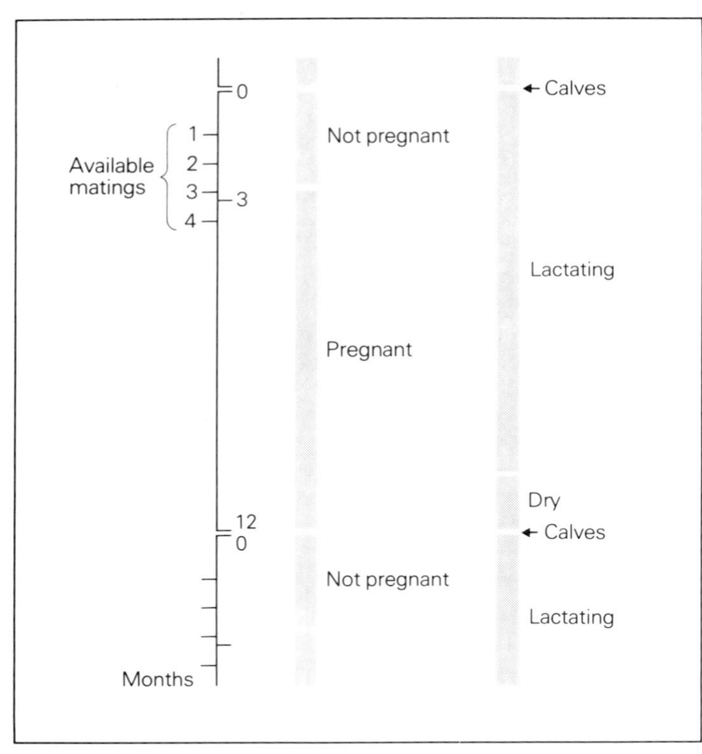


**Figure 5.10** 365-day breeding cycle for a dairy cow

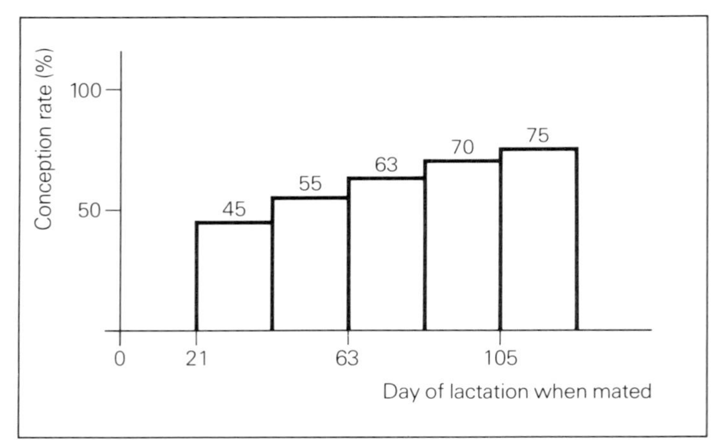


**Figure 5.11** Influence of duration of lactation on conception rate. A calving interval of 365 days is consistent with a 305-day lactation, a 56-day dry period and conception within 82 days of the beginning of lactation

apparently suffer is further exacerbated by the point in the lactation at which the cow must conceive, as this is coincidental with maximum stress when yield is highest and the metabolism working hardest.

Although cows do not show lactational anoestrus, it appears natural enough that the normal tendency would be for a high-yielding cow at peak performance to wish to forgo, temporarily at least, the delights of imposing another pregnancy upon the system. Even though lactation does not block rebreeding in the dairy cow, it is necessary to accept that with high-yielding animals maximum lactation may be irreconcilable with a 365-day calving interval and mating could be delayed until 15 weeks or more after calving.

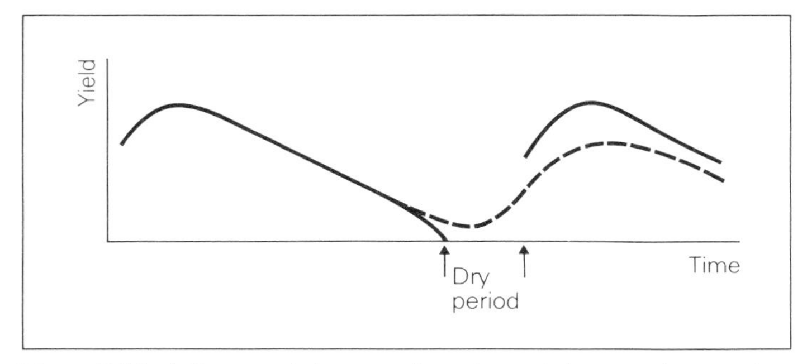


**Figure 5.12** Influence of absence of any dry period upon subsequent lactation yield. The broken line shows the result of continuous lactation with no break between the termination of one lactation and the beginning of another

The maternal body and the mammary gland both require a resting period between lactations in order to refurbish body stores and condition, and to regenerate a new secretory epithelium of milk synthesizing cells on the inner surface of the alveoli. If dairy cows have not ceased to lactate within 30–50 days of the next, impending, calving, then the end of the lactation should be imposed (Fig. 5.12). There is a relationship between milk yield and the length of the dry period which preceded it for dry periods up to 40–50 days in length. It is usually recommended that cows should be dry at least 30–40 days, but preferably for 50–60 days between lactations.

A protracted period of more than 60 days between the end of the lactation and the next calving is of course quite ruinous to productive efficiency. This would follow from either delayed conception for the next calving *not* matched by a persistent current lactation or, equally, timeous conception but a precipitous lactation decline. The length of time that cows are allowed to eat but not produce must be kept to the minimum commensurate with mammary tissue regeneration. As far as can be estimated presently, the best combination of factors is achieved with one calf per year and a lactation of high peak and good persistency lasting 300–320 days.

# Nutritional value and quality of milk

# 6

The nutritional value and essential nature of milk to the newborn human (whether the milk be sucked from mother, or requisitioned from another mammal) is indisputable. For the weaned child and adult, however, milk taken from cows for human consumption must compete with alternative sources of human nutrients. About 10 million tonnes of milk are produced in England and Wales annually. About 70 per cent goes for liquid consumption and the rest to manufacture. Of the milk used for manufacture about 30 per cent is made into butter and another 30 per cent made into cheese. Cream accounts for a further 20 per cent while the remainder goes to condensed and powdered milk.

**Table 6.1** Percentage contribution to the total allowance of nutrients made by the provision of 0.5 litre of milk

| | 5-year-old child | Adult |
|---|---|---|
| Energy | 20 | 10 |
| Protein | 30 | 20 |
| Iron | *negligible* | |
| Ca | 60 | 75 |
| Thiamine | 30 | 15 |
| Riboflavine | 80 | |
| Vitamin C | 70 | 50 |
| Vitamin D | *negligible* | |

The contribution made to the daily nutrient requirements of a 5-year-old child and an adult by the daily supply of $^1/_2$ litre of milk is shown in Table 6.1. It is a noticeable characteristic of the better fed countries of the world that it is the

**Table 6.2** Protein available from various nutrient sources (g/head/day)

| | Better fed countries | Worse fed countries |
|---|---|---|
| Cereals | 30 | 33 |
| Pulses | 4 | 12 |
| Meat | 20 | 5 |
| Milk | 20 | 3 |

consumption of meat and milk which is particularly elevated (Table 6.2).

The monetary worth of milk is difficult to judge in changing economic circumstances, but at UK prices Table 6.3 shows the competitive nature of milk as a cost-beneficial nutrient source.

**Table 6.3** Weight of protein from 100 g edible food of animal origin together with comparative values (UK, 1977)

| | g protein/100 g food | g protein purchased per unit of currency (meat = 1) |
|---|---|---|
| Meat | 15 | 100 |
| Sausage | 10 | 85 |
| Cheese | 25 | 125 |
| Eggs | 13 | 110 |
| Milk | 3.3 | 150 |

## Nutritive content

The various components of milk are shown in Fig. 6.1, while average compositions of a range of milks are given in Table 6.4.

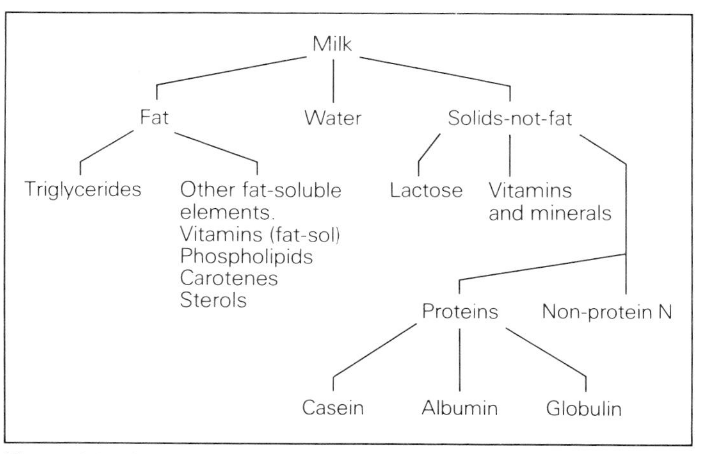


**Figure 6.1** Components of milk

It can be seen from the table that cow's milk is one of the less concentrated, tending to be low in fat, protein or lactose, or of all three, when compared to other species. Be that as it may, the nutritional value of milk in general, and of cow's milk in particular, is not open to question. Milk is 92–98 per cent digestible and the biological value of the protein is 85 to 90. Milk readily supplies energy, protein, vitamins and minerals to those that imbibe it. It is nearly self-sufficient in the ingredients required of the human diet. Government-backed

**Table 6.4** Approximate average composition of milks (%)

| | Total solids* | Fat | Protein | Lactose |
|---|---|---|---|---|
| Woman | 13 | 4.5 | 1 | 7 |
| Rabbit | 33 | 18 | 14 | 2 |
| Sheep | 17 | 6 | 5.5 | 4.5 |
| Pig | 18 | 8 | 6.0 | 4.5 |
| Goat | 13 | 4.5 | 3.5 | 4.5 |
| Horse | 11 | 2 | 3 | 6 |
| Cow | 13 | 4 | 3.5 | 5 |
| Buffalo | 18 | 8 | 5 | 5 |
| Zebu | 14.5 | 5 | 4.0 | 5 |
| Friesian | 12.4 | 3.7 | 3.4 | 5 |
| Ayrshire | 13.0 | 4.1 | 3.6 | 5 |
| Jersey | 15.0 | 5.5 | 4.0 | 5 |

*Total solids are the sum of the constituents other than water and the value is equivalent to the dry-matter content.

Solids-not-fat = total solids – fat.
Water content = 100 – total solids.
Cows milk contains about 0.7 per cent ash, 0.12 per cent Ca and 0.10 per cent P.

campaigns, common to many countries, encouraging milk consumption are soundly based on the knowledge that if 0.5 litre of milk is ingested daily, there is relatively little nutritional harm that can befall the recipient, regardless of any other nutritional insults or deprivations that might occur.

The mammalian newborn is totally dependent upon milk for nutrients. Starch digesting enzymes are absent from the tract, and although some digestion of non-casein protein may be possible within the first three weeks of life of calves and pigs, this is definitely not the case for non-lactose energy sources. Table 6.5 shows some digestibility values that one might expect from young calves or piglets. The activity of lactose-digesting enzyme is high in young mammals but tends to fall towards weaning, while the activity of protein-digesting enzymes is high for milk proteins, but is limited for non-milk proteins. Milk is rich in energy, containing about 25 MJ/kgDM. The newborn needs energy for maintenance, for creating warmth to combat a cold environment, and for growth. He has poor insulation, limited energy reserves and a high growth impulsion. The growth of mammalian young, particularly in respect of their rapid accumulation of body fat, can be judged from Table 6.6. The greater is the need of the youngster for energy, the richer the milk from the mother tends to be. Thus, species which must grow fast, accumulate large body fat reserves, and be weaned early, must have milk with high fat contents. The seal has a milk with 50 per cent fat, which is a good thing as seal pups are often neglected three weeks after birth. The rabbit has 18 per cent fat, the pig 8 per

**Table 6.5** Digestibility values expected in young calves or piglets (% digested)

| | Birth | 2 weeks old | 4 weeks old |
|---|---|---|---|
| Milk casein | 95 | 95 | 95 |
| Milk lactose | 90 | 95 | 95 |
| Starch | 0 | 25 | 50 |

**Table 6.6** Composition of milk-fed young

| | Water (%) | Fat (%) | Protein (%) | Ash (%) |
|---|---|---|---|---|
| *Piglets* | | | | |
| Birth (1.25 kg) | 81 | 1 | 11 | 4 |
| 5 kg (3–4 wks) | 68 | 12 | 13 | 3 |
| 10 kg (5–6 wks) | 66 | 15 | 14 | 3 |
| *Calves* | | | | |
| Two-day-old (33 kg) | 70 | 3 | 21 | 6 |
| 65-day-old (60 kg) | 67 | 8 | 20 | 6 |
| *Lambs* | | | | |
| Two-day-old (3 kg) | 74 | 1 | 20 | 5 |
| 50-day-old (13 kg) | 67 | 9 | 19 | 4 |

cent and the cow only 4 per cent. The energy in cow's milk can be estimated by use of the equation (adapted from the Tyrrell–Reid equation) given below.

Energy in cow's milk (MJ/kg)=0.39 (%BF)+0.2(%SNF)—0.24

Where %BF is the percentage of milk fat and %SNF the percentage of solids-not-fat (total solids—fat) in the milk.

About 35 per cent of milk dry matter is the highest quality protein. This is used by the sucking young to replenish protein losses resultant from the high rate of tissue protein turnover shown by young animals, and for growing new tissue, particularly lean body mass in the form of muscle. In the young, 50–70 per cent of growth is as lean tissue, which contains around 22 per cent protein and 78 per cent water. The gross efficiency of conversion of milk protein to animal body protein is as high as 70 per cent. A breakdown of the nitrogen-containing components of cow's milk is shown in Table 6.7, while the amino acid composition of the protein is given in Table 6.8. Most milk protein is in the form of casein (80 per cent of the protein), only about 13 per cent of the protein being as albumin and globulin. Examination of the amino acid spectrum shows whole milk protein to be unsurpassed for purposes of growing muscle protein, the composition of which is shown in the right-hand column of Table 6.8.

**Table 6.7** Breakdown of protein components of cows milk (%)

| | | | |
|---|---|---|---|
| 5 non-protein nitrogen | | | |
| 95 proteins | casein | 80 | |
| | milk serum proteins | 13 – albumins | 9.5 |
| | | – globulins | 3.5 |

**Table 6.8** Amino acid compositions expressed as percentage of the total protein

| Amino acid | Whole milk protein | Whole egg protein | Wheat protein | Muscle protein |
|---|---|---|---|---|
| Isoleucine | 6.5 | 6.9 | 4.0 | 4.7 |
| Leucine | 9.9 | 9.4 | 6.0 | 8.0 |
| Lysine | 8.0 | 6.9 | 3.0 | 8.5 |
| Phenylalanine | 5.1 | 5.8 | 4.0 | 4.5 |
| Threonine | 4.7 | 5.0 | 2.8 | 4.6 |
| Tryptophan | 1.3 | 1.6 | 1.2 | 1.1 |
| Valine | 6.7 | 7.4 | 4.0 | 5.5 |
| Methionine | 2.4 | 3.3 | 1.3 | 2.5 |
| Cystine | 0.9 | 2.3 | 2.0 | 1.4 |

Milk is a physiological fluid and is mostly water. The relationship between water-soluble milk constituents and total yield is fairly static; most changes in milk total solids result from changes in milk fat. Those changes in the solids-not-fat fraction (lactose+protein+minerals) which do occur are mainly attributable to changes in protein as a result of the cattle getting insufficient dietary energy (dietary *energy*; it will be recalled from Chapter 2 that synthesis of milk protein requires energy).

Another characteristic of milk is its content of immune bodies in the form of the immunoglobulins. The human foetus can receive immune bodies across the placenta, but this does not occur in the pig or cow. Immunity transfer by the milk during the first few hours of life is therefore vital to the survival of the newborn mammal. It is primarily the first milk of the cow which is rich in immune bodies, and it is only for the first few hours of life that the digestive systems of the young are able to allow those protein bodies into the system without being broken down. The protein level in cow's milk falls from about 14 per cent at calving, to 10 per cent 6 hours later and back to the more normal 4 per cent by 1 day after the birth. The extra protein in the milk is as immune bodies. Immune bodies comprise about 60 per cent of the protein in the first milk and about 2.5 per cent of the protein in milk thereafter. Most of the milk immunoglobulins immediately after calving are absorbed intact into the body of the newborn suckler. From the second day onwards, the situation changes and most of the immunoglobulins are not absorbed into the body of the sucking young, but have potent antibacterial activity at the surface of the gut lining. One of the problems of early weaning is that this antibacterial activity is lost, and the newly weaned young are particularly susceptible to intestinal infections.

Strictly speaking, the young mammal can only be weaned when there is no longer any specific requirement for milk. This will occur when the young have a patent enzyme system for non-milk nutrients, when active immune mechanisms are established within the body of the young which is then self-sufficient against disease, and when solid food can be consumed in sufficient quantities for the animal to survive and grow. In addition, it would not be sensible to wean unless alternative food sources were available, and there was an economic necessity to either rebreed the mother (of anoestral species) or draw off the milk for other purposes.

## Factors affecting compositional quality

The total solids content of milk from dairy cows can readily vary from 11 to 14 per cent, the solids-not-fat fraction from 8 to 9 per cent and the fat fraction from 3 to 5 per cent. One of the most significant factors affecting milk composition is the breed of the cow concerned; the influence of breed is shown in Table 6.9. Individual cows under the same management can also commonly show differences at the extremes of the

**Table 6.9** Compositional quality of milk of different breeds

| | Yield (kg) | Fat (%) | Total solids (%) |
|---|---|---|---|
| Ayrshire | 4100 | 3.9 | 12.6 |
| Dairy shorthorn | 3900 | 3.6 | 12.3 |
| Friesian | 4600 | 3.7 | 12.4 |
| Channel Island | 3400 | 4.8 | 13.8 |

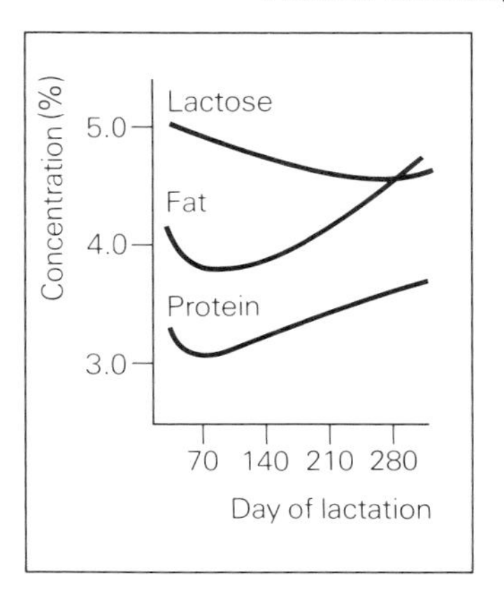


**Figure 6.2** Influence of stage of lactation upon milk composition

compositional range. In general, low fat is associated with high yield, and changes in protein content tend to go hand-in-hand with changes in fat content.

The age of the cow and the stage of lactation are two infamous causes of reduced milk quality. With each successive lactation fat declines by about 0.05 percentage units and solids-not-fat decline by about 0.1 percentage units. Some of this effect may be due to yield increasing up to the fourth lactation, but mammary troubles also increase with age and these bring about decline in both fat and solids-not-fat. At the start of the lactation the first milk has twice the normal concentration of solids, five times the protein, about twice the fat and half the lactose. After the composition has settled down 2 to 3 days after calving, a pattern such as is shown in Fig. 6.2 usually becomes apparent. Fat and protein content vary inversely to yield while lactose goes into steady decline over the whole of the lactation. Total solids-not-fat are high at first, but decline rapidly until about 6 to 8 weeks after calving. From the second month of lactation until the eighth month or so the drop in lactose is countered by the rise in protein, with the consequence that the solids-not-fat content remains steady. After the eighth month of the lactation, solids-not-fat content will tend to rise again under the influence of the pregnancy hormones. In contrast to the case of the pregnant cow, the barren animal will show a steady fall in both solids-not-fat and milk fat over the whole of the remaining part of the lactation.

Protein and lactose tend to vary little from day to day, any changes being gradual. Fat content, however, varies considerable from day to day and a contributory factor is variation in the residual milk from milking to milking. At the beginning of each milking session the fat will be lower than at the end of milking. The fat percentage can vary from 2 per cent in the fore milk to 4 per cent in mid-flow and 8 per cent in the stripping. There is little change in solids-not-fat during the course of the milking. Where the milking intervals are unequal, milk fat tends to be higher after the shorter milking interval; on a 16/8 hour milking interval routine for example, the fat content might be 3.5 per cent and 4.5 per cent. This is due to the residual milk being high in fat: as the quantity of residual milk is proportional to yield, there is more residual milk after the longer interval which lowers milk fat at that milking. As the residual fat is carried over to the next milking, the fat content of milk drawn after the shorter milking interval will be higher, but total lactation milk fat production will not be affected.

Ill-health, particularly mastitis, will affect both milk yield and milk composition. Mastitis causes a drop in milk lactose and potassium level and a rise in sodium, chlorine and the serum protein. Three-quarters of all herds whose milk has been rejected due to low solids-not-fat are likely to have had a mastitis problem; half of these will probably have had cell counts of over 1 million cells/ml (the no-problem cell count is 0.25 million cells/ml). The main effect of mastitis is upon the solid-not-fat fraction, through the reduction in lactose content.

Environmental and seasonal changes also bring about compositional changes in the milk. The comfort zone for dairy cattle is between 5° and 25° C; temperatures below or above

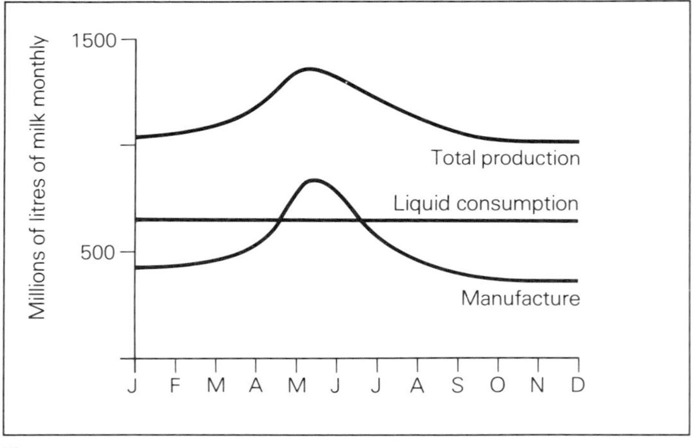


**Figure 6.3** Seasonality of UK milk production, liquid consumption and manufacture (95% of liquid milk is pasteurised; manufactured milk includes cream, butter, cheese, condensed milk and milk powder)

will reduce yield. Low temperatures may increase fat and solids-not-fat content, while high temperatures are usually associated with a general decline in both milk fat and lactose.

Seasonal trends for total milk production in UK are shown in Fig. 6.3. The spring peak results partly from the national herd having its busiest calving time in winter and early spring, but the main effect is likely to be due to the great improvement in nutrition which takes place when the cattle are fed on spring

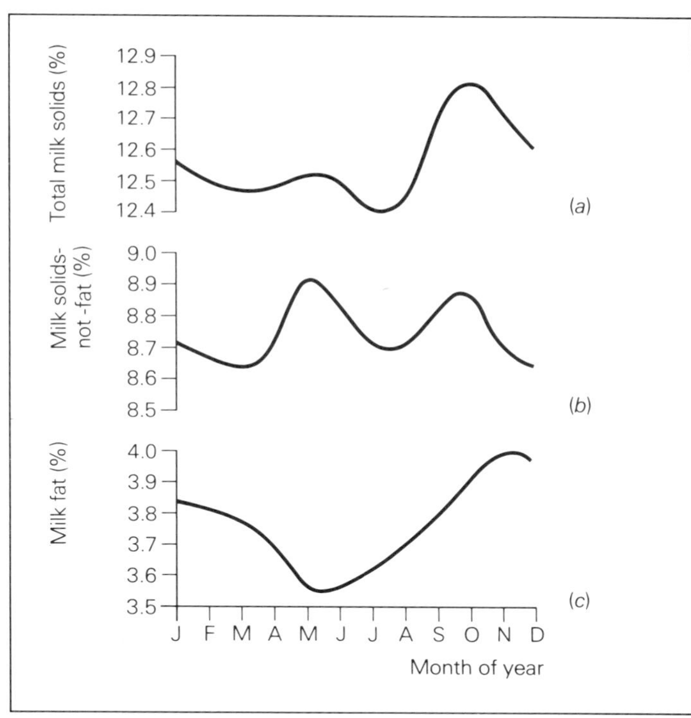


**Figure 6.4** Influence of month of year upon (*a*) total milk solids, (*b*) solids-not-fat, and (*c*) milk fat percentage in UK

grass and have ample feed supplies. Milk fat tends to be high in winter and low in spring (Fig. 6.4*c*); this may be partly a result of stage of lactation, fat being inversely related to yield, but is probably mostly due to the level of fibre intake which is likely to be higher on winter diets than on spring grazing. Winter reductions in solids-not-fat (Fig, 6.4*b*) are usually due to a shortfall in energy supply on winter diets, particularly when the roughage fraction of the diet is not of the highest quality. Solids-not-fat tends to rise again when cows go out to grass in the spring. This reflects the improvement in energy nutrition which takes place in the spring season of the year.

## Nutrition and milk composition

Cows in poor condition due to underfeeding in the previous lactation and during the immediate pre-lactation dry period are particularly prone to produce milk that is low in both milk fat and solids-not-fat concentration. Poor nutrition during the course of the lactation has similar consequences. An increase in energy supply to the lactating animal either by an increase in food consumed or by provision of a more concentrated diet will bring about a tendency to increase yield, to improve the protein (and thereby the solids-not-fat) percentage and to reduce the fat content of the milk. The fat is reduced on two accounts; firstly because of the inverse relationship that milk fat has with milk yield, and secondly because of another inverse relationship between milk fat content and the supply of concentrate feedstuff. Short-term changes in energy intake during the lactation tend to affect yield rather quicker than they affect milk fat. Temporary reductions in yield therefore can cause milk fat content (expressed as a percentage) to increase, and conversely an increase in yield will reduce fat percentage. In general, however, the overall level of nutrition has a much greater influence on milk yield than on the percentage of fat in the milk. It is prolonged underfeeding which adversely affects fat percentage the most, the extent of the reduction in fat depending on body reserves.

Variations in solids-not-fat are mostly attributable to variations in the milk protein fraction. Protein is little affected by protein intake itself, depression only occurring in cases of severe protein under-feeding. It is the energy content of the diet which affects the protein fraction the most. A 40 per cent reduction in energy can result in an 0.4 percentage unit drop in solids-not-fat. Where solids-not-fat rise with spring turn-out it is as a result of an improvement in energy nutrition effecting a rise in the milk protein level.

As the proportion of roughage in the diet decreases or if that roughage is ground, then the milk fat content will decrease. Diets containing equal amounts of energy, but in which concentrates have replaced roughages, show a depression in milk fat (the adverse effects of high concentrate diets are exacerbated if the concentrates contain a high proportion of maize). The influence of dietary constituents upon milk fat have been explained in terms of their relative ability in producing acetic or propionic volatile fatty acids from rumen fermentation processes. The relationship is not however a simple one. In general, the higher the proportion of concentrates in the diet the lower the proportion of acetate yielded from the rumen as energy substrate. The main precursor of milk fat is, of course, acetate; so a low rumen yield of acetate will reduce milk fat. Fine grinding of roughages has a similar negative effect upon milk fat concentration on account of this treatment causing an increase in feed intake and a reduction in acetate yield from the rumen. Heat treatment of feeds and pelleting the diet, in common with chopping and grinding the roughage fraction, are all feedstuff processes likely to depress acetate production and increase propionate yield. (Conversely, it has been recently suggested that the addition of sodium bicarbonate to the diet can increase rumen alkalinity and thereby decrease propionate production.)

Dairy cattle concentrate diets require a normal level of about 2–4 per cent of fat in them. Higher levels of dietary fat, particularly of saturated fatty acids (usually from tallow and oil cake sources), can increase the fat content of milk, but high dietary levels of above 6 per cent might conceivably interfere with rumen functions.

The lactose fraction of the solids-not-fat is synthesized from blood glucose, and is insensitive to variations in blood glucose within the normal range of 40–80 mg/100 ml. Variations in diet do not therefore greatly affect milk lactose.

In practical conditions milk quality problems tend to be chronic, insidious and intransigent. It is difficult to increase milk protein content other than by improving cow body condition and feeding level to a significant extent; or if the problem is due to low lactose from mastitis, by eradicating the disease. Such actions as these would require a complete reappraisal of the whole farm policy. Low milk-fat levels can often be associated, by one means or another, with high yields and one would be loath to change that phenomenon. In the short term, however, it may be necessary to forgo some extra yield that would have resulted from a higher level of concentrate feeding in order to keep the roughage intake up to maintain fat levels in the milk. One of the solutions to this side of the problem is to use, for dairy cows only, the highest quality roughage. This can encourage cows to produce both a high level of milk and also result in a high level of acetate production to keep the milk fat level up. Consistent production of high quality roughage is also, unfortunately, a long-term farm-management problem rather than one which could be solved instantaneously. Where a milk quality problem remains after the nutritional factors have been accounted for as far as possible within the circumstances prevailing, changing the breed, or strain within breed by use of a selected bull, is probably the only long-term solution.

## Payments for quantity and quality

The purchaser may pay the producer for his milk according to quantity alone, or to quantity adjusted for quality. Premium payments for milk may relate to the content of milk-fat, milk protein or total milk solids. Occasionally supplementary payments may be made on the strength of a known high class market for milk of particular characteristics, as would be the case for Jersey cow milk in the UK. Milk purchasing schemes differ widely between areas of production, and within any area according to demand and fashion (whether, for example. milk-

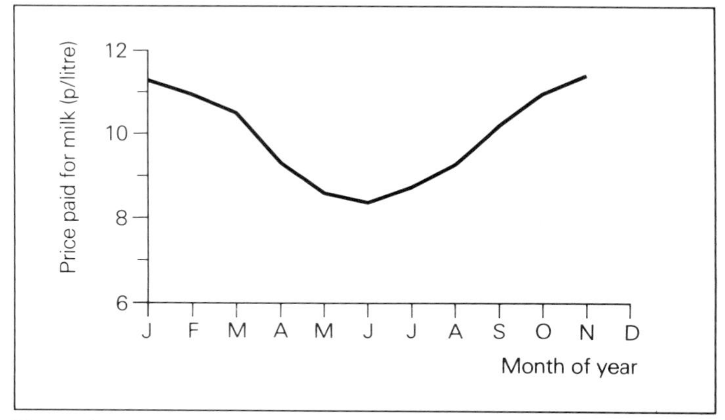


**Figure 6.5** Seasonal trends in prices paid to milk producers (UK, 1978)

**Table 6.10** Adjustments to average milk price according to milk quality (Scotland, 1978)

| Percentage of total solids in the milk* | Adjustment made (p/litre) | Percentage of producers |
|---|---|---|
| >13.2 | +0.36 | 5 |
| >13.0 | +0.30 | 8 |
| >12.8 | +0.24 | 15 |
| >12.6 | +0.18 | 21 |
| >12.4 | +0.12 | 22 |
| >12.2 | +0.06 | 16 |
| 12.0–12.2 | 0.00 | 9 |
| <12.0 | –0.10 | 3 |
| <11.8 | –0.20 | 1 |
| <11.5 | –0.60 | <1 |

*Minimum solids-not-fat 8.4 per cent, minimum milk fat 3.0 per cent.

fat is taken to have positive or negative value).

An examination of one area of the UK shows that in Scotland in 1977 the Scottish Milk Marketing Board purchased about $1060 \times 10^6$ litres of milk. Some $570 \times 10^6$ litres of that were sold for liquid consumption and $490 \times 10^6$ litres diverted to manufacture. In UK as a whole, there is a strong seasonal trend in milk production but not in liquid milk consumption (Fig. 6.3), thus the milk manufacturing trades must deal with the severe handicap of an uneven supply of raw material. The value of milk to the manufacturer is much lower than that to the consumer of liquid milk; the average price paid to producers must therefore reflect the proportion of the total production going to manufacture (Fig. 6.5). In Scotland, the price paid to milk producers is adjusted according to the total solids contents of the milk as given by the scales laid out in Table 6.10.

# Tailpiece

Any sort of an attempt to glimpse at where we go next can only be subjective, but is perhaps worth a little speculation on the basis of what has gone before. It is a self-evident truth in the agricultural industry that the short-term future will see increasing herd size and increasing cow productivity. What is perhaps worth a little deeper thought is the role of the buffalo and Zebu in non-temperate climates; it is here that great potential lies, not simply in improving native breeds and producing crosses with higher yields, but in devising milking methods and routines which are particularly suitable for these cow types.

The oxytocin story is far from complete. One wonders sometimes whether the absolute, 'acute', implosion reaction is *quite* so absolute. There are variations in response between species, between animals and within individual animals which need explaining. Particularly, there is the possibility of oxytocin acting in a more 'chronic' way by continuous secretion into the blood during milk removal, or perhaps by multiple, but discrete, doses. The reproductive problems which are beginning to plague high-output production systems might also benefit from a better fundamental understanding of oxytocic hormone activity, as this is also involved in ejaculative and orgasmic responses.

The milking machine has far to go before it approaches the satisfactory standards set by a sucking calf; whether this can be achieved best by continuing efforts to improve current designs, or by considering radically different designs is a matter of conjecture. The technology for increasing the efficiency of the milking routine is clearly with us. One hundred and twenty cows milked by one man in an hour is perfectly feasible with a static or rotary parlour which incorporates features of automatic cow control (for entry to and exit from the parlour), automatic udder washing before milking, automatic udder spraying with disinfectant after milking, which omits examination of the fore milk, and which removes the cluster mechanically when milk flow has stopped. All the operative has to do is to identify the cow for purposes of computerized recording, dry her after automatic washing, check that the teats are not diseased or disabled and apply the teat cups of the machine.

An interesting conundrum is raised by the shape of the lactation curve. Is the quest for ever higher peaks to be pursued *ad infinitum*, or should one now be looking for cows with greater lactation persistency and higher mid-lactation yield? In any event, how many of those apparently biologically constant features of milk production are really a result of man's impositions, rather than of any natural laws?

# Further reading

**Austin C. R. and Short R. V.** (Eds.) (1972–1976) *Reproduction in Mammals*, (5 volumes), Cambridge University Press, Cambridge.

**Cowie A. T. and Tindall J. S.** (1971) *The Physiology of Lactation*, Edward Arnold, London.

**Mepham B.** (1976) *The Secretion of Milk*, Edward Arnold, London.

**N.I.R.D.** (1977) *Machine Milking*, National Institute for Research in Dairying, Shinfield, UK.

# Index

accumulated milk yield, 54
acetate, as precursor of milk, 20, 84
adrenalin in milk production, 28, 34
age at first calving, 65
age of cow, and milk quality, 81
air phase, of milking machine, 46
albumin, in milk, 79
alveolus of mammary gland, 10, 11, 15, 18, 19
alveolar,
  collapse time, 37
  contraction, 37
  epithelium, 13
  growth, 22
  implosion, 36
  lumen, 19
  tissue, 14, 15
amino acid composition,
  of egg, 79
  of milk, 79
  of muscle, 79
  of wheat, 79
amino acids, as precursors of milk, 19–20
annual yield, 73
annular tissue of mammary gland.14, 15
anoestrus, 24, 74
antibacterial activity of milk, 80
apocrine secretion, 20
appearance of udder, external, 6, 8
appetite, and yield, 71
automatic cluster removal devices, 53–4, 58, 61, 62
automatic teat spraying, 60
autumn calvers, and lactation yield, 66–7
Ayrshire, 63, 78, 81

behavioural responses, 33, 34, 39
biological value of milk, 77–80
blood,
  glucose as precursor of milk, 19–20, 84
  supply to mammary gland, 16–17
body stores, and milk yield, 71
body-weight changes, 71
breeding cycle, 72–5
breeds of dairy cattle, 1, 80
  and compositional quality of milk, 80
  and yields, 63
buffalo, 1, 9, 51, 78
butter, 3, 4, 76
butting, in suckling behaviour, 36, 39–40; *see also* nosing

calf, feeding of, 29–39
calving,
  interval and lactation, 66, 72–5
  month of, and lactation, 66–8
casein in milk, 19, 20, 21, 79
cell activity in mammary gland, 23, 24, 64, 67
cell counts for milk, 55, 82
cell losses in milk, 20, 55, 82
cell numbers in mammary glands, 12, 64, 67

cell turnover, 55
central nervous system and the milking process, 16, 27–8
cessation,
  of lactation, 24–6, 75
  of milk removal, 12, 72
Channel Island, 63, 81
cheese, 3, 76
cistern of mammary gland, 9, 10, 11, 14, 15, 29, 35
  size, 34–5, 51
    in relation to yield, 56
cluster,
  design, 47
  removal, 62
    automatic, 53–4, 58, 61, 62
  weight, 47, 52
collapsed liner time, 47
composition of milk, 19–20, 77–85
  effect of nutrition on, 83–5
  factors affecting, 80–3
  from various breeds, 81
  from various species, 78
composition of milk-fed young, 78, 79
concentrates, feeding of, in parlour, 57, 58
conception, delay in, 75
conception rate, and lactation, 73, 74–5
condensed milk, 76
conditioned stimuli, 34, 50
condition of animal, and lactation yield, 65, 71
constituents of milk, *see* composition of milk
consumption of milk, 3–4, 78
contraception and lactation, 24
cow numbers, 1–2
cream, 3, 76

daily yield, 65, 68; *see also* lactation curves
Dairy Shorthorn, 63, 81
damage to tissues, 20, 46, 50
decline of lactation curve, 68, 69, 70, 73
demand feeding, 30, 31, 39
development of mammary gland, 9–13
digestibility of milk, 77, 78
discharge of milk, 19–22, 26–8
discomfort, and its effect upon milk removal, 34
distribution industry, 1–4
dry period, 26, 68, 69, 74, 75
  importance of length of, 72
ducts of mammary gland, 9, 10, 11, 14, 15, 18
  growth of, 9, 22

early lactation, 23, 68–9
early weaning, disadvantages of, 80
eccrine secretion, 20
efficiency of conversion of milk, 79
ejection of milk, 26–8, 34, 35–9, 51, 54
ejection reflex, 27, 35–9, 49–52
elasticity of mammary gland, 14
embryonic growth of mammary gland, 9–10
energy content of milk, 78, 79
energy intake and milk composition, 84
energy reserves, of young, 78
energy utilisation, in milk synthesis, 21
enzymes and the newborn, 78, 80
epithelium of mammary gland, 18–20, 23, 24, 26
  regression of, 26
equipment for milking, 44, 56–62

fast milking cows, 46, 47, 51, 57
fat,
  droplets, in secretion of milk, 20
  in diet, 84
  in milk, 19, 20, 77, 81–5
    depression of, 84–5
    of different breeds, 80–1
    effect of season on, 83
    influence of roughage on, 85
    influence of yield, 81, 83, 84
    low levels of, 85
    and unequal milking intervals, 82
  reserves, the need to accumulate, 78
fatty acids,
  in diet, 84
  in milk, 20, 21
fatty tissue of mammary glands, 15
feed supply for lactating cow, 71–2
flow, of milk, 35–8, 46–9
  cessation of, 53
  and machine pulsation, 47–9
  rates of, 54
foremilk, 58
forequarters, output from, *v.* rear, 54
frequency of milking, 23, 26, 29–31, 39, 56
frequency of suckling, 30, 31, 56
Friesian, 2, 63, 78, 81

genetic selection, 63
gland cistern, *see* cistern
globulin, in milk, 77, 79
glucose, precursor of lactose, 20
glycerol, in formation of milk fat, 20
goat milk,
  composition of, 78
  production of, 1, 2
grass, and its effect on lactation curve, 66
growth,
  of mammary gland, 9–13, 22
    between lactations, 13
  of young, 78–9
grunting, in relation to milk flow, 40, 41

half-life of oxytocin, 32, 33
hand milking, compared with suckling, 34
heat treatment of feeds, 84
heifer lactation, 70

herd size in Europe, 2
herringbone parlour, 57, 59
hormonal control of lactation, 18, 22–8
horse milk, composition of, 78
hose rinse, 50, 51, 58, 62
human breast feeding, 30, 37
hygiene in parlour, 55, 61, 62
hypothalamus, role of, in milk production, 28, 33, 35

ill health and milk yield/composition, 82
immunoglobulins in milk, 80
implosion of alveoli, 27, 35
ineffective milking, 54
inguinal canal, 16, 17
infection of mammae, 55
infertility of dairy cows, 73–4
initiation of lactation, 33–4
inspection of equipment, 62
interval,
 between calvings, 66, 72–5
 between milkings, 20, 56
intramammary pressure, 52
intestinal infection of newly weaned young, 80
irregular milking, 72

Jersey, 78, 85

lactation,
 beginning of, 22, 25
 cessation, 24
 characteristics of, 63
 curves, 64–71
  peak height of, 70, 71
  calculation of, 69
 cycle, 25
 decline of, 12, 23, 24, 25
 early, 23, 68–9
 end of, 24
 inhibition of, 25
 initiation of, 33–4
 late, 68–9
 length of, 65, 68, 70, 72–5
 mid, 68–9
 number, 66
 peak of, 12, 23, 68, 69, 70
 yield, 23, 63–75
lactose,
 in milk, 19, 20, 77, 80, 81, 82
 synthesis, 22
lambs, composition of milk-fed, 79
late-lactation, 68–9
lateral suspensory ligaments, 14, 17
lean body mass of young, 79
liner of teat cup, 42, 43, 47
liner collapse phase, 42
liquid milk, production and consumption of, 1, 2–4, 76–8 *passim*, 82, 86
litter size, effect of, on yield, 64
lobes of mammary gland, 11, 14, 15
lobule of mammary gland, 14, 15
loss of yield, due to mastitis, 55

machine,
 pulsation, 47–9
 removal, 53, 62
 stripping, 52–5
machine-milking, 42–62
machine-on time, 51–4, 58
maintenance of equipment, 62
mammary gland (mamma), 5–28
 blood supply to, 16–17
 cell activity in, 23, 24, 64, 67
 cistern of, 9, 10, 11, 14, 15, 29, 35
 damage to, by machine, 49
 development of, 9–13
 ducts of, 9, 10, 11, 14, 15, 18
 elasticity of, 14
 embryonic growth of, 9–10
 epithelium of, 18–20, 23, 24, 26
 fatty tissue of, 15
 hormonal control of, 22–8
 lobes of, 11, 14, 15
 lobules of, 14, 15
 microscopic structure of, 18–19
 nerve supply to, 16, 17
 pressure in, 20
 secretion in, 19–21
 shape of, 5–7, 8
 stretching of, 6, 14
 suspension of, 14, 17
manpower requirements, 56–60
manual massage, 56
manufacture of milk, 4, 76, 82, 86
massage,
 by operator, 50, 52
 by young, 34, 39
mastitis, 20, 42, 53, 55–6, 82, 85
mating and lactation, 74, 75
maximising yield, 70
medial suspensory ligaments, 14
microscopic structure of secretory tissue, 18
mid-lactation, 69
milk,
 as a food, 76–7
 constituents, *see* composition of milk
 ejection, 26–8, 34, 35–9, 51, 54
 flow, 35–8, 46–9
  cessation of, 53
  rates of, 54
 industry, 1–4
 losses due to mastitis, 55
 manufacture, 4, 76, 82, 86
 monetary worth of, 77, 85
 production, 2, 76
 products, 1, 3
 release, 8–9, 33, 34
 removal, 23–4
 retention, 9
 secretion, 19–22
 storage, 29
 synthesis, 12–13, 14, 19–22, 23
 value, as food, 4, 76–80
 yield per suckling, 36, 37
milking,
 frequency, 23, 26, 29–31, 39, 56
 ineffective, 39

interval, 20, 29, 56
machine, 42–62
parlour, 45, 56–62
phase (*vac*) of milking machine, 42, 43
rate, 46, 50
routine, 60
stalls, in parlours, 57, 58, 59, 60
termination of, 39
times, 53, 57, 60
minerals in milk, 20, 77
month of calving and lactation, 67
mouth movements of young, 37, 40
multiple young and milk release from mammae, 38
muscle,
action of sphincters, 16, 27, 34, 35, 39
cells of mammary gland, 18
convulsion and oxytocin, 33
myoepithelium, myoepithelial cells, 18, 19, 26, 27, 28, 29–32, 33, 34, 37
contraction, 26, 31–2, 33
contraction time of, 37

national milk production, 2, 82
natural milking, 29–41
nerve supply to mammary gland, 7, 16–17
nervous system, role of, in lactation, 16, 18, 27, 28, 32, 33, 34, 50
neurohormonal milk ejection reflex, 27, 28
nipple, 11, *see also* teat
nosing in suckling behaviour, 39; *see also* butting
nursing behaviour of pigs, 39–41
nutrition,
influence of, on udder size, 11
of lactating cow, 66, 71–2, 82
requirements of child and adult, 76
nutritional value of milk, 76–80

oestrogen in gland development, 10. 22
oestrous cycle, 10, 72
operation time in parlour, 56–8, 60, 62
ovary, 22
overmilking, 54
oxytocin, 23, 24, 26, 27, 28, 32, 33, 34, 35, 36, 37, 38, 40, 51, 52
activity of, 38, 51
catabolism of, 33
doses, multiples of, 38
release of, 38, 40

paraventricular nucleus (PVN), 28
parlour, milking, 28, 42, 45, 56–62
herringbone, 45, 57, 59
rotary, 60
routine, 7, 9, 32, 33, 38, 39, 51, 56–62
faults, 55
throughput, 60
passive withdrawal, 34–5, 36, 37, 40, 51, 54
pattern of milk flow, 36, 37
payment for milk, 85–6
peak yield, 65
and three times daily milking, 56
time of, 64
persistency of lactation, 69, 70, 71, 75
phases of sucking by young, 36, 39
pig,
composition of milk, 78
nursing behaviour, 31, 39–41
pituitary gland, rôle of, in lactation, 22, 28, 32, 35
placental hormones, 22
of mammary gland, 9–10
post-natal development of mammary gland, 9–10, 12
post-pubertal growth, 10, 22–4
powdered milk, 76
precursors of, 19–20
predicted lactation curve, 69
pregnancy, 11, 12, 22, 24, 25, 72
antagonistic to lactation, 24
fifth month of, 68, 70
hormones, 81
influence on lactation length, 72, 73
premium payment on milk quality, 85–6
prenatal development of mammary gland, 9–10
pre-partum milking, 22
pressure in gland, 46, 56
production of milk, 2, 76
progesterone, 11, 22, 24
prolactin, 11, 22, 23, 24
propionate, 84
protein,
in milk, 19, 20, 77, 79, 80, 81, 83, 84
turnover, 79
puberty, 10, 12
pulsation of milking machine, 44, 47–9

quality of milk, 76, 80–3
problems, 85
quality payment, 85, 86
quiescent period, 26

rabbit milk, composition of, 78
reabsorption, 26, 72
rearquarters, output from *v.* fore, 54
receiver jar, 44, 58, 62
recovery phase (*air*) of milking machine, 42, 43, 46
reflex response, 33
refractory period after milking, 38
regeneration of mammary tissue, 12, 26, 75
replacement of cells, 24
reproductive cycle, 22
residual milk, 39, 47, 52, 82
retention of milk, 14, 34
rotary parlour, 60
roughage,
in diet, 84
processing, 84
routine, in milking parlour, 7, 9, 32, 33, 38, 39, 51, 55–62

rhythmic grunting by suckling sow, 40, 41

seal milk, composition of, 78
season of year and its effects,
 on lactation, 66, 67
 on milk composition, 82, 83, 86
 on milk production, 82
 on prices, 85
secretion, 22
 inhibition, 26
 of milk, 11, 19
 rate, 52, 56, 71
secretory,
 cells, 24
 epithelium, 12
 tissue, 63
 premature loss, 55
senility, 24
shape of gland, 5–7, 8
shape of lactation curve, factors influencing, 70–1
sheep, 1
 milk, composition of, 78
sinuses, 9, 14, 15
slow milking, 55, 57, 58
slow sucking, 41
smooth muscle in mammary gland, 29, 32, 33
solids-not-fat, in milk, 19, 79, 80, 83
sow, 35, 39
sphincter, in mammary gland, 14, 16, 32, 33, 35, 36, 42, 46
spray washing of udders, 58
spring, effect of lactation curve, 66
spring calver, 66, 67
spring hump, 66
stage of lactation and its effect on milk composition, 81
stalls, 57, 58, 59, 60
starch, digestibility of, 78
sterilisation of equipment, 62
storage of milk in gland, 7, 14, 19, 20, 23, 39, 56
streak canal, 10, 15
stress and its effect on yield, 28
stripping, 52–5
stripping weight, 52
sub-clinical mastitis, 55
substrates for milk, 20
sucking, 22, 26, 33, 35, 39
 interval, 29
 mechanism, 34
 stimulus, 24, 26
suckler cow, 64
suckling, 34–41
 behaviour, 39–41
 unsuccessful, 41
supplementary payments for milk, 85–6
support tissues, 7
suspensory ligaments, 7, 14, 15, 17
synthesis of milk, 12–13, 14, 19–22, 23, 71, 72
synthesising cells, 5
synthetic activity of cells, 24
tactile stimulation, 50
teat,
 cluster, 52
 cistern, 14, 15
 cup, 42, 43
 crawl, 49
 dipping, 58, 60, 61
 disfiguration, 42
 orifice, 14, 46
 size of, 54
 recovery, 49
 shape, 6, 7
 sphincter, 15, 47
 strength of, 54
 structure, 54
temperature and composition of milk, 82
termination of lactation, 69, 72
three times daily milking, 56
tissue damage, 42
total solids in milk, 80, 83, 86
 in different breeds, 81
trigger mechanism, 38
triglycerides, 20

udder, *see* mammary gland
underfeeding, 84
under-milking, 54
uterine muscle, 22
uterus, 24
utilisation of milk, 3, 4, 86

vac phase, of milking machine, 42, 43, 44, 46, 47, 48, 49
vacuum,
 characteristics, 58
 fluctuation, 55
 level, 47, 48, 49, 50, 55
 pressure, 43
 stability, 49
value of milk as food, 4, 76–80
vitamins, in milk, 20

washing, 50, 58
water in milk, 20, 80
weight loss, 71
weight,
 of animal, 65, 66
 of cows, 63
 of young, effects of, on yield, 64
winter calver, 66, 67

yield,
 annual, 2
 daily, 65, 68
 decline, 64, 65, 69, 72
 different breeds, 81
 influence of nutrition, 84
 lactation, 63
 maximization, 70
 peak of, 65
 time of, 64
 per cow, 1, 2
 persistency, 65
 potential, 2
 relationship with young, 64
 of sows, 64
Zebu, 1, 9, 29, 52, 78